5500
80C

SEPTIC SYSTEMS HANDBOOK

SECOND EDITION

O. BENJAMIN KAPLAN

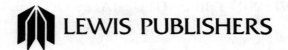

LEWIS PUBLISHERS

Library of Congress Cataloging-in-Publication Data

Kaplan, O. Benjamin.
   Septic systems handbook / O. Benjamin Kaplan—2nd ed.
   p.   cm.
   Includes bibliographical references and index
   1. Septic tanks.   2. Water, Underground.   3. Soils—
Leaching.
   I. Title.
   TD778.K36   1991   90–42891
   628.7′42—dc20
   ISBN 0–87371-236-6

LEWIS PUBLISHERS, INC.
121 South Main Street, Chelsea, Michigan 48118

PRINTED IN THE UNITED STATES OF AMERICA

This book is dedicated
to I. Halmos,
who taught me
to be truthful to myself

O. Benjamin Kaplan earned his BS in Soil Science from California State Polytechnic University, and his MS and PhD in the same field from the University of California at Davis. In 1973, Dr. Kaplan obtained an MS in Public Health from the University of California at Los Angeles. He was hired as a registered sanitarian by San Bernardino County, California, in 1974. His subsequent publications, lectures, and activities in the environmental health field earned him listings in *Who's Who in Health Care* (1977) and *Who's Who in the West* (1978).

As a county sanitarian, Dr. Kaplan applied his dual educational background to the improvement of local practices in the field of septic systems. His experiences in this area form the backbone of this book.

Dr. Kaplan ended his public service career in January, 1987, and is currently working as an environmental health consultant.

Preface

Over the last decade, I have reviewed over 3,000 percolation test reports (and septic system designs) prepared by a variety of individuals (mostly civil engineers, and some engineering geologists, sanitarians, and architects). From those reports—and conversations with the preparers—I found that none had received academic training on how to evaluate soils and design site-suitable septic systems. Their reference manuals (USPHS Publication 526 and the Uniform Plumbing Code) contained significant inaccuracies and were rather obsolete. These same problems bothered other colleagues who reviewed percolation test reports for other California counties. Also, most land planners I have talked with are aware of some of the environmental impacts caused by sewering an area, but few are aware of the profound land use consequences of development dependent on septic systems.

So, I set out to write a concise but comprehensive book to emphasize basic theory and yet be practical for everyday use by all types of professionals who deal with septic system design, use, and approval (particularly those who work as septic systems consultants). This book is the result.

O. B. Kaplan
1986

Foreword to the Second Edition

The first edition of this book emphasized technical knowledge; it has been incorporated into the second edition with
subtle but important modifications. However, professional
practice does not depend only on technical knowledge, and
it does not take place in a vacuum. It is affected by legal,
economic, and ethical considerations. The ways in which
this occurs are explored in Appendices M through T. Appendix U offers tips on solving some problems commonly
encountered in professional practice.

O. B. Kaplan
1990

Acknowledgments

This second edition has benefited from insights provided by R. B. Brown, Lou and Doreen Blanck, Barry and Joyce Eskin, Scott Maass, Marlin Fernandez, Jon Lewis, and Robin Berry. I am also thankful to my wife, Adele, who helped to condense the text.

Contents

xi

List of Figures

List of Tables

SEPTIC SYSTEMS
HANDBOOK

Why Public Health Agencies Control the Disposal of Domestic Sewage

I used to believe that everyone knew that sewage poses a health hazard. After all, hygiene is taught at home and in prep school. Epidemiology itself was born as a science in the 1600s, when Snow related cholera epidemics to sewage contamination of groundwater. At the UCLA School of Public Health, I learned from Professor C. Senn that, until pit privies were introduced, millions of people in the South were parasitized by hookworms and, too weak to work, were considered to be "lazy." World Health Organization bulletins from Senn's bookshelves spoke of elevated morbidity and mortality due to improperly disposed sewage; one of them described an occasion when introduction of piped-in, clean potable water to a developing-country village increased morbidity and infant mortality because no measures were taken to dispose of the water after it was used.

So, I was quite surprised and shocked when I heard an "expert" testify in front of a local planning commission that sewage was not a health hazard.[1] Later on, I met more than a few such "experts."

Therefore, it seems proper to start this book by giving solid reasons why the disposal of wastewater or sewage must be controlled by public health agencies. For our purposes, sewage is any domestic wastewater, be it "blackwater," which drains down the toilet, or "greywater," which drains from all other plumbing fixtures (handbasin, kitchen sink, tub, shower, etc.).

1.1 CONSEQUENCES OF IMPROPER DISPOSAL OF SEWAGE

Ponded sewage near inhabited areas may afflict residents in various ways. As a nuisance, it may generate offensive odors, it may attract rodent pests, and it may serve as a breeding ground for mosquitoes and flies. Obviously, such conditions do not help the neighborhood's property values. As a health hazard, sewage may contain parasitic worms' eggs and larvae, and also microbial pathogens and parasites. Some of these may attack man directly through the skin, or after transmission by a vector (usually rodent or insect), or after man ingests sewage-contaminated food or water. Of the "top five" human parasitic diseases,[2] each with about half a million to a million cases per year worldwide (ascariasis, hookworm, malaria, trichuriasis, and amoebiasis), only one (malaria) is not directly spread in sewage. A wealth of pertinent, detailed, and authoritative information can be found elsewhere.[3,4] A brief mention of diseases propagated through direct or indirect contact with sewage and of their causative agents follows. (See Appendix A for summary.)

1.2 CATEGORIES OF CAUSATIVE AGENTS AND SPECIFIC DISEASES

Various Types of Tapeworms, Roundworms, and Flatworms

Ancylostomiasis. The hookworm larvae penetrate the skin of the feet and travel to the gut.

Ascariasis. The roundworm eggs stay in the sewage-contaminated soil; after they are ingested (dirty hands, contaminated food), they develop in the gut. The adults may attack lungs, liver, and other organs.

Dracontiasis. These unusual roundworm larvae are shed from the skin with the washwater (greywater), and are ingested by a tiny aquatic "bug" (Cyclops); the roundworm infects people who drink (the bug in) the water.

Enterobiasis. The adult female roundworm, "pinworm," injects its eggs near the anus of the victim. The eggs are easily spread in sewage-contaminated irrigation waters, and contaminate leafy vegetables.

Strongyloidiasis. The roundworm larvae in contaminated soil penetrate the skin of the feet and move to the lungs and gut.

Somatic cysticercosis. The eggs of the tapeworm are ingested with contaminated water, hatch in the gut, and then the larvae may attack various organs: eye, brain, heart.

Schistosomiasis. The eggs of the flatworms are discharged in the urine or feces; the larvae grow inside aquatic snails, are discharged by them, swim to and penetrate the skin of people who might be wading nearby, and grow to adulthood in the veins of the victims.

Trichuriasis. Eggs of the roundworm develop into embryos in contaminated soil; after the embryos are ingested (dirty food, hands), they grow in the gut.

Yeast

Candidiasis. The yeast is transmitted by contact with feces or secretions from infected people. Although it causes usually mild infections, occasionally it may cause ulcers in the intestinal tract, or lesions in the kidneys, brain, or other organs.

Protozoa

Amoebiasis, balantidiasis, and giardiasis. The protozoa or their cysts are transmitted through contaminated water, contaminated raw vegetables, and flies; they attack the gut and cause mild to severe diarrhea.

Bacteria

Cholera, salmonellosis, shigellosis, and typhoid fever. The bacterial agents may be ingested in food or water contaminated by sewage; some may be transmitted by flies which have contacted feces.

Viruses

Epidemic and sporadic viral gastroenteritis, hepatitis A, and polio. The viral agents are transmitted in sewage-contaminated water or food.

Others

Allegedly, there are infectious agents even smaller than viruses which are called prions, and which are suspected of causing some uncommon degenerative diseases. Someday, one or another kind of prion or an unknown type of life form might be found to be spread through contaminated food or water. We should keep our minds open to such possibilities.

Although sewage is a potential health hazard, some people may experience direct or indirect contact with sewage and suffer no consequences. Often the sewage may not contain many dangerous parasites and pathogens. But, if nowadays few people discharge such parasites .and pathogens in their sewage, that's because public health agencies have implemented programs of immunization and of provision of potable water and sewers (or septic systems).

1.3 THE COST OF PREVENTION

One may ask, if the risk derived from improper sewage disposal is low because few people discharge pathogens or parasites in their sewage, why not allow a few substandard septic systems? As economists would say it, the marginal benefit of demanding acceptable standards up to the very last septic system in a community may be lower than the marginal cost of doing so. In most human endeavors, there is a point of diminishing returns, beyond which the marginal (incremental) benefit/cost ratio is unfavorable. It just doesn't pay to put more money or effort beyond the point of diminishing returns.

However, things are somewhat different in the field of Public Health, as noted elsewhere.[5] The nature of this difference is explained below by means of two hypothetical events: a farmer's attempt to maximize his profit by fertilizing his crop up to the point of diminishing returns, and a public health administrator's attempt to maximize the health of some villagers.

Table 1.1 presents the hypothetical crop production operation. As shown, if a farmer increases fertilizer applications from 0 to 0.15 tons/acre, he will spend $150 per acre but will gain back $200 per acre. So, he is ahead $50 per acre. The benefit/cost of this increase in fertilizer is 1.3, which is larger than 1 and hence favorable. If the farmer increases fertilizer applications from 0.15 to 0.30 tons/acre, he will spend $150 per acre and will get back $150 per acre; the benefit/cost is equal to 1, neutral (he gets back as much as he puts in). If he goes beyond this point and applies more than 0.30 tons/acre, he will start losing money.

The common economic schedule thus far presented is not

Table 1.1 Derivation of Benefit/Cost Ratios and Determination of Most Profitable Fertilizer Application Schedule, on a Per Acre Basis

Tons of fertilizer	0.00	0.15	0.30	0.45	0.60
Resulting crop yield, tons	1	3	3.5	4	4.2
Net revenue from crop (at $100 per ton), $	100	300	350	400	420
Marginal revenue, $		200	150	50	20
Net cost of fertilizer (at $1000 per ton), $	0	150	300	450	600
Marginal cost of fertilizer, $		150	150	150	150
Marginal or incremental benefit/cost ratios		1.3	1.0	0.33	0.13

quite applicable to prevention of communicable disease, as illustrated below.

Let us assume that a transient who carries a communicable disease spends a week in a village with 100 inhabitants. During that week he transmits his disease, directly or indirectly, to two villagers. The transient leaves, but in the course of the second week each of the infected villagers transmits the disease to two other villagers, and so on, every succeeding week. At first the progress of the disease is very rapid (logarithmic progression); but later it slows down in proportion to the number of people already affected, as shown in Table 1.2 and in Figure 1.1.

Now, medical intervention in the first week would require treating only 2 people and would protect 98 people. But, if postponed to the fifth week, it would require treating 95 people and would protect only 5. The magnitude of the costs involved are quite different, not to mention the suffering.

Therefore, it is not difficult to surmise that, when dealing with a communicable disease:

1. It pays to intervene as soon as possible. Prevention is similar to intervention on or before "week number zero."
2. Prevention is generally more desirable and can be more cost-effective than cure.

Table 1.2   The Spreading of a Chronic Nonlethal Communicable Disease over a Population, at a Rate of Two Random Direct or Indirect Transmissions Per Infected Person Per Week

Week No.	Number of New Cases	Cumulative Number of Cases
1	2	2
2	3.9	5.9
3	11.1	17.1
4	28.3	45.4
5	49.6	95
6	9.7	100+

Note: Transmissions are random, and some people already have the disease when it is transmitted to them. Therefore, the new cases column is computed by multiplying the previous week's number of cumulative cases times two, and times the probability that the prospective victims are still free of the disease. For instance, during the third week, $5.9 \times 2 \times (100-5.9)/100$ equals 11.1, and 11.1 plus 5.9 equals 17.1, neglecting rounding off errors.

PERCENT

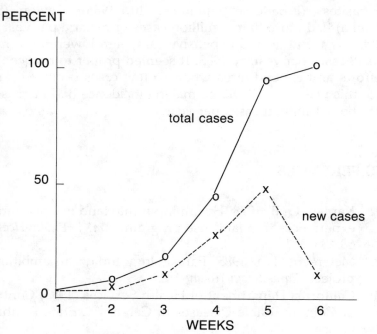

Figure 1.1   Spread of disease; percent of population affected.

3. It is generally wise to allow no exceptions to prophylactic measures. It takes only one person to start the "infection snowball" rolling.

1.4 RELEVANCE

The hypothetical illustrations in the previous section are not mere illusions.

Mathematically speaking, the parallelism between the spread of a communicable disease and the spread of an imported agricultural pest is striking. The pest affects areas instead of villagers. Prevention assumes the form of a quarantine of certain agricultural products at state or U.S. borders. Also, decisive action is taken the moment an infestation is discovered.

According to a PBS documentary entitled "Conquest of the Parasites" (broadcast September 10, 1985), Ceylon (Sri Lanka) had more than 1 million cases of malaria per year in 1980. A vigorous and expensive campaign lowered the rate to 18 cases per year by 1983. It seemed proper to reduce the efforts and expenditures when so few cases occurred, and so, this was done. By 1985, malaria incidence had increased to about 1 million cases per year.

REFERENCES

1. Kaplan, O. B. 1978. Health input into land use planning: experiences in a land use program. *Am. J. Public Health* 68:489–491.
2. McGregor, Ian. 1985. Parasitology today: an ambitious project. *Parasitology Today* 1:2.
3. California Department of Health Services (1983) Control of Communicable Diseases in California. State Printing Office.

4. Benenson, A. (Ed.). 1970. *Control of Communicable Diseases in Man*. American Public Health Association, Washington, DC.
5. Kaplan, O. B., and El-Ahraf, A. 1977. On the economics of justifying a preventive public health program through benefit-cost analysis: The case of restaurant inspections. *J. Food Protection* 40: 566–568.

2

The Septic System

Sewers carry sewage away from its point of origin to a faraway treatment plant or disposal area. Onsite systems treat and/or discharge sewage within the site where the sewage originates. The most common type of onsite system is the septic system, also called the septic tank sewage disposal system, or septic tank soil absorption system (ST-SAS). Approximately one-third of all houses in the United States are served by septic systems.[1] Figure 2.1 illustrates a septic system and its components: the septic tank itself, and one type of leachfield or soil absorption system, a leachline.

2.1 THE SEPTIC TANK

As seen in Figure 2.1, sewage drains out of a house through a pipe called a "house sewer," which discharges into a device called a septic tank. This tank has a liquid capacity of about 500 to 1500 gallons (sometimes higher), depending on the amount of sewage generated daily. The tank may have various shapes (Figure 2.2) and may be made out of concrete, galvanized and coated steel, fiberglass, or polyethylene.

The tank's function is to receive sewage and to hold it for a little while. During this detention period the "floatables" in the sewage (oils, greases, and some fecal constituents) float to the top, where they undergo some microbial decomposition and form a floating layer of white-brownish scum. This

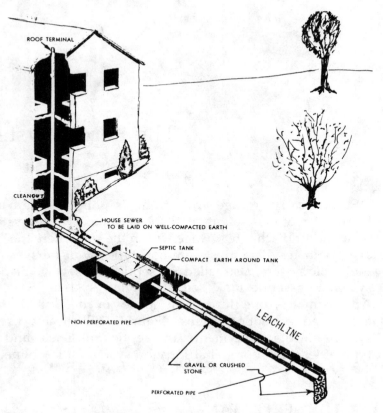

Figure 2.1 A septic system: septic tank and portion of a leachline. (Adapted from U.S. Public Health Service.[2])

detention period also allows the sewage "sinkables" to settle to the bottom of the tank, where they undergo little decomposition (due to the anaerobic environment); they acquire a black color, and become sludge. Between the layer of scum and the layer of sludge remains a translucent, greenish liquid called clarified sewage; it trickles to the leachfield when displaced by a fresh load of incoming sewage. (The tank buffers sudden incoming flows and releases the clarified sewage slowly and continuously for up to roughly 20 minutes after the incoming flows stop.) The relative positions of

Figure 2.2   Shapes of septic tanks. (Source: U.S. Public Health Service.[2])

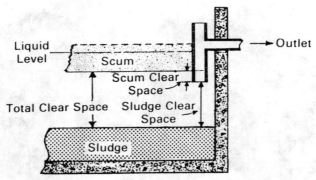

Figure 2.3 Scum and sludge within a septic tank. (Source: Otis et al.[1])

scum, sludge, and the clear space (clarified sewage) are shown in Figure 2.3.

Every tank must have vents. These are shown in Figure 2.4. One purpose of the vents is to prevent sewage flows from draining by vacuum the "u"-traps in the house plumbing. The other purpose is to allow the escape of (explosive) methane and malodorous gases from the tank. These gases are generated when (facultative) bacteria decompose some of the sewage constituents. The gases may travel from the top of the tank's liquid level into the vent, and back up to the house sewer pipe and up to the house roof terminal (Figure 2.1), where they are vented to the atmosphere. Malodorous gases may become a quite noticeable nuisance during the first year of operation of the septic tank.

A licensed septic tank pumper should pump the sludge and scum as frequently as necessary to prevent them from accumulating to such an extent that they reach the entrance to the outlet "tee" (Figures 2.3 and 2.4) and flow out of the tank into the leachfield. (The leachfield is damaged by scum or sludge particles in two ways: The particles directly plug up the leachfield's soil pores, or else they stimulate the growth of microorganisms which can plug up the soil pores and decrease infiltration. This may cause the sewage to pond within the leachfield and surface out on top of the ground.)

The frequency of pumping must depend on the rate of

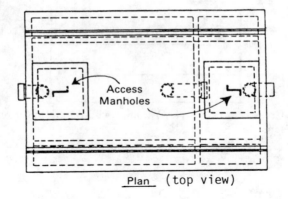

Plan (top view)

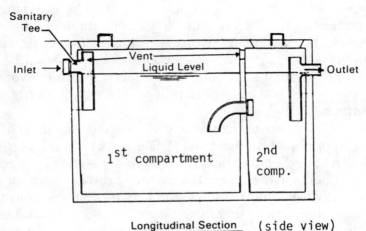

Longitudinal Section (side view)

Figure 2.4 Typical two-compartment septic tank. (Adapted from Otis et al.[1])

scum and/or sludge accumulation, and this rate in turn depends on many factors, including amount, composition, and frequency of sewage discharges. As a very rough approximation, one person generates about 3 cubic feet of scum plus sludge per year. Use of a garbage disposal in the kitchen sink increases the discharge of solids to the septic tank and the frequency of pumping. Also as a very rough approximation, the tank should be pumped every 2 to 5 years, commonly every 3. But each individual user should

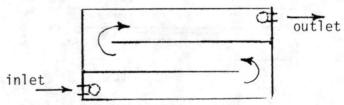

Figure 2.5   Cross section (top view) of improved-removal-efficiency septic tank.

determine his own best pumping frequency by evaluating the accumulation of scum and sludge at the time of the first pumping. (In passing, it is worth mentioning that septic tank additives—such as yeast, bacteria, and enzymes—widely sold for "digesting scum and sludge" and "avoiding expensive pumpings" have not been proven effective in controlled experiments.[1-3])

Figure 2.4 shows horizontal and vertical cross sections of a two-compartment septic tank. The sanitary "tees" at the tank's inlet and outlet and the baffle or separation between compartments help prevent the incoming sewage from moving directly from inlet to outlet, and help retain the scum and sludge. In practice, there are some sewage constituents which have almost the same density as water, and stay within the clarified sewage for a long time. Other constituents might be a bit more dense, but they pick up tiny gas bubbles from the anaerobic fermentation that occurs in the tank. Thus, they become lighter, and stay in suspension within the clarified sewage until discharged. I myself have seen a piece of toilet paper flowing out of a septic tank in a bare trickle of clarified sewage. But generally, under average conditions, only clarified sewage flows out of the tank.

Winneberger conducted extensive research on septic systems and concluded that the tank's removal efficiency could be substantially increased by lengthening the path of sewage within the tank,[4] as shown in Figure 2.5.

2.2 LEACHFIELD

The function of all types of leachfields is to receive the clarified sewage from the septic tank and discharge it underground into the soil. The soil environment purifies this liquid by breaking down its biodegradable components, and by retaining parasitic worms, their eggs, and microbes. The purified liquid moves away by percolation and evaporation, and by plant uptake and transpiration.

There are three main types of soil absorption fields: the leachline, the seepage bed, and the seepage pit.

Leachline

The leachline, as shown in Figure 2.1, is a gravel-filled trench with a perforated pipe running through its length; the trench runs below ground level. (In the past, before the advent of synthetic-material perforated pipe, gapped segments of clay pipe were used instead of perforated pipe.)

The perforated pipe is installed level and is plugged with a cap at the end, so that the liquid will drip out of every perforation along its length. In practice, the dripping occurs through the first few perforations if the pipe is perfectly horizontal. But often the pipe sags, just a little, and most of the liquid (clarified sewage) is discharged at the pipe's lowest point. Thereafter, the liquid spreads along the bottom of the leachline trench.

As the liquid spreads over the soil, over time it induces the growth of a "biomat" or "clogging mat" or "clogging layer"[4-6] on the wetted soil. This mat is composed mostly of facultative (aerobic/anaerobic) bacteria and bacterial products (a "goo" or slime of polyuronides and polysaccharides colored black by ferrous sulfide precipitates). This mat extends, from the surface of wetted soil, not more than about 1 or 2 inches into the soil. If the surface of the wetted soil is well-aerated (aerobic), a variety of soil nematodes and protozoa digest the mat's bacteria (and perhaps their "goo").

The aerobic mat is more permeable than the anaerobic mat,[7] but this is not too important, as discussed later in this book. Under very anaerobic conditions, the mat's permeability (with liquid ponded about 0.5 foot high over it) is not lower than about 0.2 gallons per square foot per day, or less commonly, half this rate.[6]

With time, the mat may extend throughout the leachline bottom, and then move up the sides of the leachline trench. Eventually, if the daily flow of sewage into the leachline exceeds the amount that can infiltrate through the mat and through whatever soil remains unclogged, the sewage surfaces over the leachfield. When this happens, the leachfield is said to have "failed." (This word is a misnomer: Leachfields do not fail; leachfield designers or installers do fail to build them to last. See Chapter 3 and Appendix B.)

The mat serves useful functions. First, it is a matrix where biological activity takes place and biodegradable materials and some microbes are consumed. Second, it filters out most pathogens and parasites. And third, it delivers liquid to the soil at a rate usually slower than the (more permeable) soil can transmit, so that the flow through the soil (percolation) is unsaturated. (Saturated flow fills every pore of the soil. Unsaturated flow fills only the smaller pores of the soil.)

Unsaturated flow enhances the soil's ability to capture in its small pores the microbes that might have passed through the mat or through the as yet unclogged soil surfaces.

Seepage bed (leachbed)

The seepage bed is another type of soil absorption system. It is merely a wide leachline (more than 3 feet in width) with one or more perforated pipes running through its length.

Seepage pit (leachpit)

The seepage pit is a vertical hole in the ground, usually about 4 to 6 feet in diameter and 10 to 40 feet in depth.

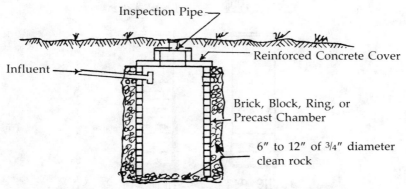

Figure 2.6   Seepage pit cross section. (Adapted from Otis et al.[1])

The vertical cross section in Figure 2.6 shows an inner supporting cylinder made of concrete bricks (or similar materials), with a layer of gravel between the bricks and the soil surface, to hold the soil in place.

(Before septic tanks became widely used, the house sewage was discharged into a [covered] vertical hole in the ground called a cesspool. Many contractors still use the word cesspool when they refer to seepage pits.)

The seepage pit works like a vertical leachline. As shown in Figure 2.7, the clogging mat moves upward along the sides, year after year; as the bottom and the sidewall near the bottom clog up, the sewage ponds at a higher and higher level. This sewage is absorbed through the (as yet unclogged) higher portions of the sidewall; also, the hydrostatic pressure forces more of the sewage through the clogged bottom and sides of the pit. In a pit, the level of liquid fluctuates up and down at least 1.8 to 4 times the extent in a leachline. (See proof in Appendix B.) This fluctuation allows a wetted soil surface to be aerated and remain more aerobic and permeable.

These two factors, hydrostatic pressure and aeration, may explain why, if absorption areas are equal and everything else is the same, a pit lasts longer than a leachline.

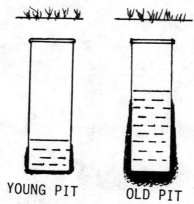

YOUNG PIT   OLD PIT

Figure 2.7   Progress of clogging mat in a seepage pit.

REFERENCES

1. Otis, R., et al. 1980. EPA Design Manual: Onsite Waste-water Treatment and Disposal Systems. EPA-625/1-80-012.
2. U.S. Public Health Service. 1967. Manual of Septic Tank Practice. PHS Pub. No. 526.
3. Kaplan, O.B. 1983. Some additives to septic tank systems may poison groundwater. *J. Environ. Health* 45: 259.
4. Winneberger, J.T. 1984. *Septic Tank Systems.* Butterworth Publishers, Stoneham, MA.
5. Otis, R. 1985. Soil clogging: Mechanisms and control. In "Proceedings of the 4th National Symposium on Individual and Small Community Sewerage Systems," ASAE Pub. 07-85, pp. 238-250. ASAE, St. Joseph, Michigan.
6. Anderson, J.L., et al. 1982. Long term acceptance rates of soils for wastewater. ASAE Pub. 01-82 (Third National Symposium), pp. 93-100.
7. Bernhart, A. 1973. *Treatment and Disposal of Waste Water from Homes by Soil Infiltration and Evapotranspiration.* University of Toronto Press, Toronto, Canada.

3

Economics of Leachfield Size

It is sometimes argued that leachline size should be reduced to a minimum: The money saved by not installing a regulation-sized leachline can be deposited in a bank to earn interest; then, if and when the reduced-size leachline fails, it can be replaced. I believe that this position is generally injudicious. My reasons follow.

1. To start with, allowing the installation of presumably short-lived leachfields runs counter to a basic aim of public health:risk prevention. (See Chapter 1.)
2. Many a developer could save money; but the leachfield user (who might not be the same developer) could be stuck with the replacement costs, and a variety of other problems:
   a. It is not easy for the average person to choose and deal with a contractor to replace a leachline.
   b. Gardens, patios, or other appurtenances can be destroyed when heavy equipment tries to reach (or excavate in) the replacement area.
   c. Heavy equipment can compact moist soil in and around the failed leachfield area, and may damage some of the pipes. The sewage-absorption capacity of compacted soil is reduced; and, as a result, the replacement leachfield may last even a shorter time than the original one.
3. The odor from a couple of leachfields that have "failed" could affect property values over an area far larger than that of the (small) lots directly involved.
4. Worst of all, leachfields that are not built to last indefinitely do not make good economic sense. (Exceptions to

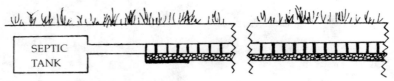

Figure 3.1   An infinite leachline.

this may exist if sewering is about to occur within a few years, or if lots are very large.)

3.1 A LOWEST-COST LEACHLINE IS BUILT TO LAST INDEFINITELY

Winneberger[1] referred to a formula for the longevity of seepage beds in some Knox County (Tennessee) soils:

Area (square feet) = 263 + 26 × years (of use before failure)

He observed that, if the bed were 400 square feet, it would last 5 years; but if the bed area were doubled, it would last 20 years—"far beyond direct proportion."[1]

The formula might not be reliable if applied to leachfields built elsewhere. But it can be shown that, anywhere, extra area indeed increases longevity far beyond direct proportion, and also that a leachline that is built never-to-fail is the most economic one. This is elaborated below.

Let us visualize an infinite-length leachline, about 3 feet wide and with only 1 inch of gravel below gapped pipe segments. For simplicity, let us assume that the rate of septic effluent discharge into this leachline is invariant, day after day. Let us also assume that all of the absorption occurs through the bottom of the leachline, and that each year about 10 feet of leachline length is clogged by biomat formation. Figure 3.1 illustrates this condition, and shows the biomat forming near the beginning of the leachline.

At the same time that clogging occurs, another process promoting or resulting in absorption through the clogged area takes place. It does not matter whether this is due to microbial decomposition of the clogging mat, to evaporation, or to residual infiltration through highly developed anaerobic mat (0.1 gallons/square foot/day), or to some other cause. Let us call the clogging process C, and the opposite, anticlogging, A. By definition, the amount of clogging is directly proportional to time, and the amount of anticlogging is directly proportional to the area already clogged. Stated otherwise, $dC/dt = c'$ and $dA/dt = c''C$, where c' and c'' are proportionality constants, t is time, and d is the differential sign. Integrating the two equalities above between time = 0 and time = t, we obtain

$$C = c't \quad \text{and} \quad A = c'' \int_0^t C dt = c''c't^2/2$$

It follows that the net length of leachline clogged, or NC, is equal to C minus A, or

$$NC = c't - c''c't^2/2$$

From this formula we can determine when net clogging will be zero. Solving for t when NC = 0, we obtain two solutions. One is trivial: t = 0 (when the leachline just begins to work, there is no clogging). The other one is $t = 2/c''$ (at this time, anticlogging consumes all of the clogged area). But, somewhere between these two times, there must be another point in time when NC reaches its maximum value, just before it begins decreasing all the way to zero. This is the time of most interest to us (for leachline design purposes), as will be seen later. To obtain the value of this particular time, we differentiate NC with respect to time, and solve for t when NC = 0. Now, $dNC/dt = c' - 2c''c't/2 = 0$. Solving for t, $t = 1/c''$. Let us now give numerical values to c' and c'' and see in a more down-to-earth way what happens to the progress of the clogging mat and to the longevity of a finite-length leachline.

Table 3.1 Linear Feet of Leachline Affected by Clogging Through Time

t, years	C, feet	A, feet	NC = C - A
1	10	0.5	9.5
2	20	2	18
3	30	4.5	25.5
4	40	8	32
5	50	12.5	37.5
6	60	18	42
7	70	24.5	45.5
8	80	32	48
9	90	40.5	49.5
→10	100	50	50
11	110	60.5	49.5
12	120	72	48
13	130	84.5	45.5
14	140	98	42
15	150	112.5	37.5
16	160	128	32
17	170	144.5	25.5
18	180	162	18
19	190	180.5	9.5
20	200	200	0

Let $c' = 10$ feet (of leachline length clogged per year), and $c'' = 0.1$ per year. The results are shown in Table 3.1.

The table shows that the net clogged area increases for 10 years to a maximum of 50 linear feet; thereafter it decreases to zero. As expected, $t = 1/c'' = 1/0.1 = 10$ years.

The NC column serves to construct a longevity table. For instance, if the leachline had not been infinite, but only 9.5 feet long instead, it would have clogged completely and failed after about 1 year of use. If 18 feet long, after about 2 years of use. If 100 feet long, it would have lasted forever.

Let us also derive the yearly cost of using leachlines of various lengths. To this end, let us assume that banks or money markets yield a net return of 5 percent on savings. Locally, leachline replacements cost roughly $200 (for moving in heavy equipment) plus $10 per linear foot. The longevity data and the costs of use are shown in Table 3.2 and in Figure 3.2.

Table 3.2 Leachline Longevity Data (from Table 3.1) and Yearly Cost of Leachline Replacement

leachline length, feet	years of use before failure	cost		$ saved	$ yield	net cost/ year
		total	yearly			
9.5	1	295	295	905	45	250
18	2	380	190	805	41	170
25.5	3	450	150	745	37	156
42	6	620	103	580	29	99
50	10	700	70	500	25	68
100	—			0	0	0
200	—			(1000)	(50)	50
300	—			(2000)	(100)	100

The first two columns in Table 3.2 come directly from Table 3.1. The total replacement cost column is obtained by multiplying the feet of leachline times $10, and adding $200; the result is divided by the years of use to obtain the next column to the right. The money saved column represents the money that was saved when the original shorter-than-100-foot leachline was installed, and is computed as 100 minus leachline length, times $10/foot. The net yield is the annual 5% return on the money saved (after accounting for inflation effects on yield and capital). The net cost per year column is obtained by subtracting yield column entries from replacement cost per year entries. The numbers in parentheses are negative values.

Figure 3.2 shows that the lowest-cost leachline is the one built to last indefinitely. If one were to build a leachline 200 feet long, which is twice as long as necessary, the yearly cost of this excessively long leachline would be the same as that of an 80-foot leachline, which is 20 feet short of the necessary length. This is indicated by the dotted line in Figure 3.2. Hence, if one isn't sure how long a leachline should be, in all likelihood it is more economical to err on the side of safety.

Socioeconomic costs include all the costs to the leachline user plus all the costs to everyone else. In addition to the

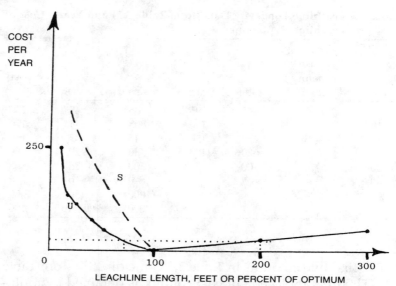

Figure 3.2 Cost of leachline use as a function of leachline length. Curve "U" represents user cost; curve "S" represents hypothetical socioeconomic costs.

costs associated with the problems mentioned at the beginning of this chapter, socioeconomic costs include those of public health agencies for purposes of enforcement and abatement of "failed" leachfields, and sewering of neighborhoods. If the leachline is larger than 100% of the area required for indefinite or long-term life, socioeconomic costs are identical to those of the user: a mere misallocation of resources—wasted money. (Therefore, the S curve is congruent to the U curve, and cannot be seen to the right of 100 feet.)

3.2 OPTIMAL SIZE OF LEACHFIELDS

The previous section gave a general view of what a septic system designer or environmental health officer should consider regarding leachfield size. Its raw data were hypothetical. The next few chapters will explain the background material needed to understand how a real leachfield is designed.

REFERENCE

1. Winneberger, J.T. 1984. *Septic Tank Systems*. Butterworth Publishers, Stoneham, MA.

4

Soils at a Glance

Four to five billion years ago, the earth's surface was covered by igneous rocks, by seas, and by an anoxic atmosphere. Heating and cooling fractured the rocks, and rainfall decomposed and leached out some of their soluble minerals. Thus, the first soil particles and soil minerals were formed from the igneous rocks. Soil particles are of three types, according to diameter: sand (2 to 0.05 mm), silt (0.05 to 0.002 mm), and clay (less than 0.002 mm). (Gravel, pebbles, and cobbles are soil texture modifiers, not soil.)

If we pulverize an igneous rock, we can obtain particles of sand, silt, and clay. The minerals in each of these three soil textural fractions are identical to those in the parent rock. But, if they are subjected to chemical weathering, new minerals appear in the clay textural fraction; they are called clay minerals (kaolinite, bentonite, montmorrillonite, etc.). Sands, silts, and clays are often extracted, segregated,concentrated, and deposited elsewhere by the erosive forces of gravity, wind, and water (including glaciers). The deposits may be fairly uniform, like wind-blown sand dunes or loess (silt) soils. They may be deposited by water streams, layer over layer of different texture. (For example, there might be a stratum of sand 3 feet thick over 2 inches of clay over 1 foot of silt, etc.)

Through time, many soil deposits are altered by a variety of agents. Chemical solutions may cement the soil particles or fill pores with new minerals. Tectonic forces may tilt nearly horizontal strata to near vertical positions, or compress the strata so strongly that the soil grains melt and almost reform igneous rock. Plant roots help break up weakened rocks, and accelerate chemical weathering. Plant decomposition products include organic acids and bases, which attack rock and soil constituents. Decomposed plants form part of organic soils, and convey special characteristics to them. Special types of soils are also created by algae (diatoms) and "animals" (foraminifera or coral deposits). Insects, earthworms, rodents, and man himself modify natural soils.

After one type of soil alteration, another may follow. As a result of all the possible combinations of (and permutations between) alteration agents, there are thousands of ways the soils of a given prospective leachfield might have formed and acquired specific characteristics.

Some knowledge of such ways is a useful diagnostic tool. For instance, in the alpine climatic region of the San Bernardino National Forest, many of the soils were formed in place by weathering of the parent rock; this rock can be found under the soils. By looking at an excavation through the soil profile down to the rock, one can estimate what kind of soil developed and what kind of problems it may present. If the rock is granitic (with minerals in the form of crystals the size of those of table sugar, and exhibiting a whitish, sparkling fresh cut), the overlaying soil may not have unusual problems. If the rock is microcrystalline (crystals are microscopic; a cut surface looks like chalk), the soil above the rock probably has a fine (clayey) texture. Drainage might be a problem, and compaction of soil may be a problem if the leachfield is installed when the soil is moist. If the rock is a pegmatite (crystals are large—about one inch), the soil might have too fast an initial percolation rate, and a much slower rate after the big pores are plugged by migrating clay or by a future clogging mat.

Some soil characteristics are fairly predictable, including a fairly common one: unpredictability. It is often difficult, impractical, or impossible to obtain all the data necessary to anticipate some complex soil behaviors.

For instance, surprising soil behaviors or conditions can be found within San Bernardino County's 20,000 square miles. In the Sleepy Hollow area I came across a groundwater depth of 6 feet in one exploratory boring, and a groundwater depth of 20 feet in another boring less than 12 feet away. (Tilted alternating strata of sandstone and shale are common in this area.) In Yucaipa I examined a reddish clay which, at a depth of 5 feet (and in my presence), consistently yielded a percolation time of 10 minutes per inch with plain tap water; and it had no structure, wormholes, root channels, or other aids to percolation. (Common percolation times for clays are well in excess of 45 minutes per inch.) An engineer* tested a sand in Morongo Valley which did not absorb water from a deep bore hole filled with over 20 feet of water. (Soil surface sands can be made impermeable by coatings of fungal products or other natural organic compounds. But deep, nonoily sands are supposed to be permeable.)

Soil permeability is a characteristic of importance to septic system designers, as it influences aeration, water flow, water retention, biological activities, and filtration of parasites and pathogens. Soil permeability is affected by texture, structure, degree of water saturation, degree of compaction, total pore space, fraction of total pore space occupied by large pores, continuity of large pores, and spatial changes in any of these variables.

Structure is the arrangement of the soil particles into shapes defined by cracks or weakness planes. These cracks may develop from expansion and contraction upon wetting and drying (particularly so if the soil has expanding type clays), compression from roots, solution channels, or weaknesses and fractures in the parent rock. If one takes a lump of soil and breaks it carefully with the fingers, one may see that it splits into smaller fragments of characteristic shapes: prisms, if prismatic structure; plates, if platiform structure; round or nutlike, if nuciform structure. If the soil grains are not structured, the "structure" is called single-grained or

*R. Carducci, personal communication.

massive. Other things being equal, the least permeable structure is the platiform. It could result from compaction and horizontal plane shear, or from residual weak planes in decomposing parent rock (like shale or slate). Nuciform structure is associated with action of fibrous roots and earthworms, and imparts high permeability even to clay soils.

Texture refers to the proportion of sand, silt, and clay in a soil.

Water saturation affects permeability in three major ways. First, it swells expanding clays, and may even disperse some clays and weaken the structural stability of a soil. Another way is relevant to permeability testing: soil that is only partially wet may have air "bubbles" in some of its pores. This entrapped air acts as a plug to water flow, until it is dissolved in the flowing water. The third way is indirect: a bit of water (just enough to change the color of dry soil) takes the soil to the point of optimal moisture for compaction. At this point the soil becomes very susceptible to compaction by tools or heavy equipment. And compaction smudges soil absorption surfaces and decreases the soil's porosity and the proportion of macropores.

Figure 4.1 shows pore size distribution in a loam soil. A vast range of pore sizes are represented in this log-normal distribution.

Halving the diameter of a soil pore decreases water flow to less than one-fourth. On the basis of cross-sectional area of flow, the decrease should be exactly one-fourth. But the sides of the pore exert a drag on the flow. As the diameter decreases, the ratio of pore perimeter to pore area increases. Hence, the drag increases, and the flow decreases more than in proportion to cross-sectional area of flow. A few big pores (macropores) can carry more water than hundreds of small pores (micropores). Macropores are visible to the naked eye. A soil might have well-developed macropores and still exhibit low permeability. For instance, a net of macropore channels might be interrupted by a very thin horizontal layer of dense silt or clay.

This brief chapter could not possibly discuss even a frac-

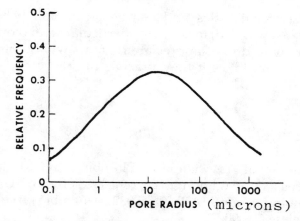

Figure 4.1 Log-normal distribution of pore sizes in a surface loam soil with well-developed structure.

tion of what is known about soil genesis, biology, chemistry, and physics. The bibliography at the end of this chapter is better suited to this purpose. Instead, this book will concentrate on soil conditions relevant to septic systems. These will be discussed as they appear in subsequent chapters.

But, even with the little information presented so far, one could debunk some simplistic beliefs held by too many people. One such belief is that soil texture suffices to determine the absorption area requirement of a leachfield. This belief is the basis for the Uniform Plumbing Code's Table I-4.[1] More will be said about this point in Chapter 15.

Another belief is that natural soils, with their cracks and macropores, are superior to artificially created soils (or fills). In fact, previous research about filtration of microbes in sieved and packed soil columns has been criticized: Those packed soil columns lack macropores and can retain the microbes far more efficiently than the natural soil.[2]

It is also important to emphasize that soils in the field cannot be treated in the abstract. Spangler and Handy[3] noted that, when faced with a problem, engineers tend to simplify and abstract, while engineering geologists tend to emphasize peculiarities and complexities. Soil behaviors regarding sewage disposal are particular and complex. Engineers who do not appreciate this fact will be prone to make major mistakes if they practice in the field of septic systems.

REFERENCES

1. International Association of Plumbing and Mechanical Officials. (1976 edition, to current 1985 edition.) Uniform Plumbing Code. IAPMO, 5032 Alhambra Ave., Los Angeles, CA 90032.
2. Smith, M.S., et al. 1985. Transport of E. coli through intact and disturbed soil columns. *J. Environ. Qual.* 14: 87–91.
3. Spangler, M. and R. Handy. 1982. *Soil Engineering* (4th ed.). Harper & Row, New York, NY.

BIBLIOGRAPHY

Written for the general public, yet quite informative, are the USDA yearbooks, *Soils and Men* (1938), and *Soil* (1957). The latter is widely available in public libraries.

Singer, M. J. and D. N. Munns. 1987. *Soils: An Introduction.* Macmillan Publishing Co., Inc., New York, NY.

5

Soil Water Movement

It is not uncommon to find someone who will design a leachfield without taking into account the peculiarities of the soil or site. For instance, on one occasion I rejected a consultant's percolation test report. In this report he recommended that leachlines be used in a prospective subdivision tract, although the tract's soils were only about 2 feet thick over solid impermeable rock. After a little while, he resubmitted the report with a new recommendation: Use seepage pits instead of leachlines.

It is also not uncommon to find someone who will misapply a general formula to a particular condition. Since one must "see" and visualize in order to understand something well, pertinent concepts and formulas will be derived and explained graphically.

5.1 INFILTRATION AND PERCOLATION

Water infiltrates into a surface, and percolates through a porous medium under the influence of gravity. The principles governing infiltration into flow-restricting surfaces are very similar, whether the surfaces belong to a clogging mat, to a smeared or compacted soil surface, or to the bottom and sides of a tin can with tiny perforations. Let us take a look at these principles.

Let us take a hypothetical case of an open, long tin can, with a thin layer of soil at the bottom. The bottom has lots of

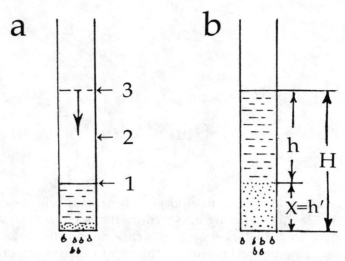

Figure 5.1 Illustration of Darcy's formula with a cross section of a tin can with perforated bottom. Dots represent water-saturated soil.

little perforations. The can is full of water and is suspended in space. Now let us try to predict the speed of the falling water level between levels 3 and 1, as shown in Figure 5.1(a). Intuitively, we would assume that the rate of fall is 3 times faster when the water is at level 3 than when it is at level 1. The pressure exerted by water is called hydraulic head, H, and can be expressed in various types of units: atmospheres, kilopascals, or feet of water. For our purposes, it is sufficient to remember that 1 atmosphere is equivalent to the pressure exerted by a column of water about 10 meters or 34 feet (in height).

Now let us go back to the can with the perforated bottom. Let us keep the water level constant at level 3, and then let us make this soil layer 2, 10, or 100 times thicker, until it reaches level 1, as shown in Figure 5.1(b). What effect does the thickening of the soil layer have on the flow of water? Intuitively we know the resistance to flow will increase 2, 10, or 100 times. So, the rate of flow will decrease, respectively, to 1/2, 1/10, or 1/100 of the original rate. It should be noted that H is the pressure exerted by the whole water column, that is, the part that is above the wet soil called surface head h, plus the part that goes through the soil to the bottom of the can, called soil head h', and equal in magnitude to the height or thickness of the wet soil, X.

Henry Darcy did similar experiments in the mid-1800s, with (real) sand filter beds, and also found out that the rate of flow is directly proportional to H and inversely proportional to X. His famous formula for water flow through porous media is

$$Q/S = k \, (h + X)/X$$

where
Q = the flow rate in cubic feet per day
S = the surface or bottom area of the "can" (or sand filter) in square feet
k = a proportionality constant reflecting the permeability of a particular soil, feet per day
h = the surface head, or height of water above the soil, feet
X = the distance of water travel through soil, feet
(X is numerically identical to h' only when X is vertical)

Note that when we divide Q by S we get units of feet per day. This is the velocity of a volume of water that goes in and out of a chunk of soil through unit surface areas. The actual velocity of water within the soil pores is higher.

From Darcy's formula, we can determine the value of a soil's k by measuring Q and S when $(h + X)/X = 1$. The term $(h + h')/X$ is called gradient; it is equal to 1 when $h = 0$ and $h' = X$. A device to measure k "the old fashioned way" is shown in Figure 5.2.

Once we find out the k of a given soil, what happens to the flow rate (the ratio Q/S) if we tilt the tube so that it makes an angle Θ with the horizontal? (Later on, the answer to this question will allow us to solve problems involving uncon-

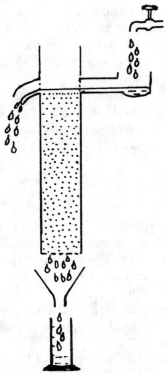

Figure 5.2   Measurement of k. An undisturbed soil sample in a tube with a perforated bottom (or wire mesh) is filled to the top (of the soil) with water. The water level is kept constant and just barely above the soil, so that h = 0. The water is collected in a graduated cylinder. The water is left to run for a while to displace all air from the soil, and to saturate every soil pore. After this saturation, the volume of water that collects in the graduated cylinder in a given time is Q. The area of the cylinder bottom is S. And k = Q/S.

fined flow.) Well, k is a property of the soil in the test tube, and doesn't change with tilt; X is the length of the wet soil in the tube, and it is still the same. S doesn't change either. The only thing that changes is H, as explained below.

As shown in Figure 5.3, the value of H is the difference in height between the top and bottom of the water column (or

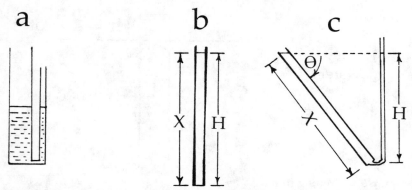

Figure 5.3 The schematic to the left, a, shows a big water-filled container connected to a thin tube. The connection allows free water flow, so the water level in the thin tube always shows the level in the container. The schematic in the middle, b, shows a test tube used to measure k under a unit gradient (H = X, so H/X or gradient equals 1). The schematic to the right, c, shows the test tube tilted at an angle Θ re the horizon; the water level in the thin tube shows the magnitude of H in the tilted tube.

wet soil) in the tube. If we know the values of X and Θ, we can determine the value of H because the sine of Θ is equal to H/X, and the sine of any angle can be found in trigonometric tables. So, now we see that the flow in such a tilted tube is given by a new expression of Darcy's formula,

$$Q/S = k \sin\Theta$$

Finally, let's do one more trick with Darcy's formula. In our imagination, let us ram a long hollow tube downward (vertically) into a dry soil, without damaging the soil or the tube. The cross-sectional area of the tube is 1 square foot, or unit area. The tube's top is 1.1 feet above the surface of the soil; the tube's bottom is 50 feet below ground. And let the soil within the tube be perfectly uniform in all directions. Now we fill the top of the tube with water until we have a surface head of 1 foot, and we keep it at that level by adding water as it is absorbed by the soil. If we measure the rate of flow of

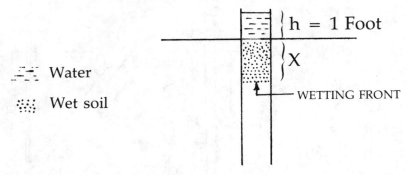

Water

Wet soil

Figure 5.4 Wetting front.

this added (or absorbed) water, we see that the water is absorbed extremely fast the first second, more slowly after 1 minute, even more slowly after one hour . . . etc. Why? This slowing-down can be explained with Darcy's equation. Please refer to Figure 5.4.

Flow velocity per unit area is equal to k(h + X)/X. At the beginning, the instant water is introduced into the tube, h = 1 foot, but X = 0 feet or pretty close to it. Therefore, the velocity of flow is extremely high. (k times 1/0 = ∞.) As time goes by and water percolates downward, the value of X increases, and therefore the ratio (h + X)/X decreases, and so does the velocity of flow. After perhaps a month or two, when X reaches a value of 49 feet, the velocity of flow reaches a value of k(1 + 49)/49, or 1.02 k. If the tube went infinitely deeper and we waited an infinite number of years until X was infinite, the velocity of flow would be stable and exactly equal to k. The length of the wetted soil, X, is measured from the top of the soil down to (just above) the wetting front.

When water percolates through soil, the wetted area is much darker than the dry soil around it. The edge of the wetted area is quite distinct, and is called "wetting front."

Although the wetting front looks like a sharp line, close inspection often reveals that the line is jagged and advances with small jerky motions, as shown in Figure 5.5. The reason

one instant

next instant

Figure 5.5 Movement of wetting front in a soil with macropores. The wet soil is above the dark horizontal line. One instant, protrusions of water move down a macropore and stop; the next instant, the adjacent small pores or capillaries fill. Thereafter the horizontal black line which represents the boundary of saturated soil re-forms at the bottom of the protrusions and the process repeats itself.

is that, in a medium-textured soil, most of the water moves through big channels — macropores; but, as soon as a macropore fills up, little channels around it suck up its water until they themselves fill up. After they do, the wetting front can keep on moving. More is said in the next section, on capillary flow.

5.2   CAPILLARY FLOW

So far we have dealt with water flowing mostly downward under the effect of gravity and pressure head. As soil texture becomes finer, going from sands to loams to silts to clays, the proportion of small pores increases; and these pores or capillaries can carry water in any direction, including upward, under the effect of tension head.

To visualize how water moves in the capillaries, let us perform imaginary experiments with easily available household implements. Let us take two glass tubes of different diameters, say 5 mm and 1 mm. Eye-droppers would work fine. Let us touch a water surface with the large- and the small-diameter tips, as shown in Figure 5.6(a). We see that water rises more in the finer tube, as if pulled up by suction.

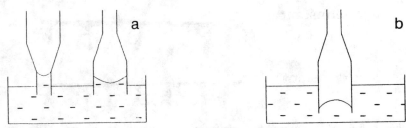

Figure 5.6 Water in capillary tubes: (a) hydrophilic tubes (b) hydrophobic tube.

Also, we see that the top of the raised water is concave; this concave surface of water is called the meniscus. If we repeat the experiment with a tube made of (or impregnated with) a hydrophobic material like candle wax or beeswax, we see that the meniscus is convex, and is "pushed" below the free water surface. We might even feel that the water is pushing the tube up, as if the tube were pressing down on an elastic membrane.

The reasons for what we see are as follows. Water molecules, like little magnets, attract each other, and are attracted by some (hydrophilic) materials like glass. The walls of the glass tube attract some water molecules upward, and these in turn "drag" other molecules upward with them. So, the water starts rising within the glass tube, and it rises until the weight of the water column is too great to be supported by the attraction between the water molecules in the meniscus' surface, or between the water molecules and glass. If the diameter of the glass tube starts increasing, the circular line of contact (perimeter = $2\pi r$) between the meniscus and the glass increases in proportion to the radius. (This line of contact can be visualized as a ring of water molecules under tension — tension exerted upward by the dry glass and downward by the weight of the water column.) But the weight of the water column increases in proportion to the square of the radius (area = πr^2). The net result is that the height of the water within a tube is inversely proportional to the ratio of perimeter to area, or H = c/r. H is the maximum height of the water column; c is a constant which depends on the strength of attraction between the tube and the water molecules, and also between the water molecules themselves; and r is the radius of the tube.

Figure 5.6(b) shows the effect of repulsion between the tube material and the water molecules; the constant c is negative, and so is H. (The line of contact is repelled or pushed down by the tube, and the hydrostatic pressure pushes the middle of the meniscus up.)

The meniscus surface consists of densely packed water molecules (which attract each other and other water molecules nearby). The same is true of the surface of a raindrop, or of a soap bubble, or of a pond on which some insects (like water skimmers) stand without sinking. Sometimes the literature refers to this molecular membrane as being due to "surface tension." It is not. If we pull both ends of a piece of wire, the wire is in tension, and it will break when the cohesion between its molecules is barely exceeded by the tension caused by pulling. Water acts in the same way. The cohesion or attraction of the molecules at the water surface is measured by pulling to the point of rupture, and the tension measured at this point is surface tension.

Now let us lift the two tubes in Figure 5.6(a), and connect the two tips that hold water. Water will not move from one to the other. If we expel some or all of the water from the thin tip, and connect this tip with the broad tip, the thin tip will "suck up" water from the broad tip. But if we reverse the procedure, the broad empty tip will not pull water out of the full thin tip. (See Figure 5.7.) Water will move from a tube with a low capillary pull head H to one with a high pull H, until the pulls are the same. A high capillary pull head H is more negative. To make things easier, we can designate capillary pull or capillary-induced tension with the letter T, for tension. This way, $T = -H$.

The rate of water movement up a capillary tube is relatively fast at first, but it becomes slower and slower as the water rises and the weight of the water column and the drag increase. On occasion, the literature refers to capillary "suc-

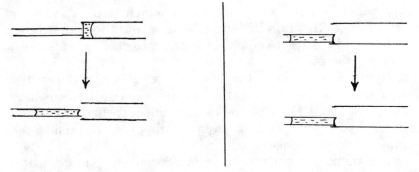

Figure 5.7 Water movement between capillary tubes.

tion," rather than "pull." The word suction is incorrect. Suction is the result of vacuum. In theory, a perfect vacuum of 0 atmospheres would cause a column of water to rise 10 meters (34 feet) at sea level, where the air pressure is 1 atmosphere. In practice, the column cannot rise more than about 8 meters because it starts vaporizing under the tension caused by the vacuum. Capillary "pull" raises water 100 meters in some tall trees. The relationship between H and T is shown in Figure 5.8.

Figure 5.8 shows water moving from a free water surface (where H = 0 and T = 0), through about 20 inches of wick, to a point 2.5 inches below the 0 level, and then dripping out. The head H is the difference in elevation (or pressure) between the free water surface and the exit point, respectively, 0 and 2.5 inches, that is, equal to 2.5 inches. The path X is 20 inches. Therefore, the gradient is 2.5/20. The drip flow per unit time is proportional to the cross-sectional area of the wick (perpendicular to flow) times a constant k for this particular wick times 2.5/20.* Now, if the dripping end is raised 2.5 inches to level 0, the flow ceases. The gradient is 0, and the values of T in the ascending and descending portions of the wick are those shown by dotted lines.

Another important point about capillary rise is shown in

---

*This is correct if the wick obeys Darcy's law. If flow occurs through continuous macropores, deviations from Darcy's law may occur.

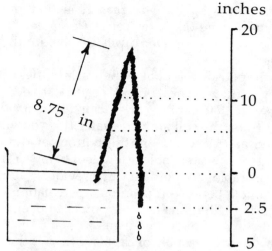

Figure 5.8 A black wick with one end submerged in water and dripping at the other end. The scale on the right is in inches.

Figure 5.9. A tube has a diameter capable of raising water 20 inches from a free water surface, but it has a bulge which itself can raise water only 0.1 inches. Water stops rising when it gets up to the bulge. But if water is introduced from above until the bulge fills up, the water can continue rising the full 20 inches.

The point illustrated above has important implications for leachfields located a few feet above a groundwater table. One implication is that capillary flow in soils does not raise water as far as theoretically possible. For instance, a soil with an average capillary diameter capable of raising water 10 feet

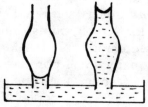

Figure 5.9 Effect of a bulge.

may raise it only 2 feet. The reason is that most of the smaller-than-average-diameter capillaries meet "bulges" or bigger-than-average-diameter capillaries before they reach 2 feet in height.

Another implication is that, when a leachfield just starts discharging effluent, the "bulges" can be filled with flow from above so that capillary rise can reach its theoretical maximum; thus, the soil and the water flow downward can be saturated all the way from the bottom of the leachfield down to a groundwater table a few feet below. (This condition is undesirable. As pointed out previously, unsaturated flow is desirable because it promotes percolation of sewage effluent through the smaller soil pores, and enhances the trapping of sewage microorganisms.)

Another important aspect of capillary flow is that the more unsaturated the soil, the slower the flow. Figure 5.10 shows that when the soil dries out a bit, soil moisture tension increases and k decreases. The soils in this figure vary in texture from Type I, coarse-textured sand (high proportion of big pores), to Type IV, fine-textured clay (high proportion of small pores).

When the Type I soil is saturated with water, its k is about 550 cm/day. (This corresponds to a Q of 135 gallons per day per square foot of area perpendicular to flow.) When it is just a bit drier, and the big pores are empty and the smaller pores still retain water at a tension T of 50 cm, k is a mere 0.15 cm/day. To give an idea of this, we could allow a chunk of unsaturated sand to dry out until it develops a tension of about 50 to 60 cm. Then, if placed just on top of a free water surface, it would pull water up to a theoretical maximum of 50 to 60 cm, if it had no "bulges."

When the Type IV soil is saturated with water, its k is about 2.7 cm/day. Its few big pores drain fast with a bit of tension, but then more and more tension is required to pull water out from its many small pores, so that the k doesn't decrease in value as fast as in a Type I soil. In fact, if the tension were about 1,000 cm, k could still be 0.1 cm/day.

It must be remembered that the curves in Figure 5.10 are

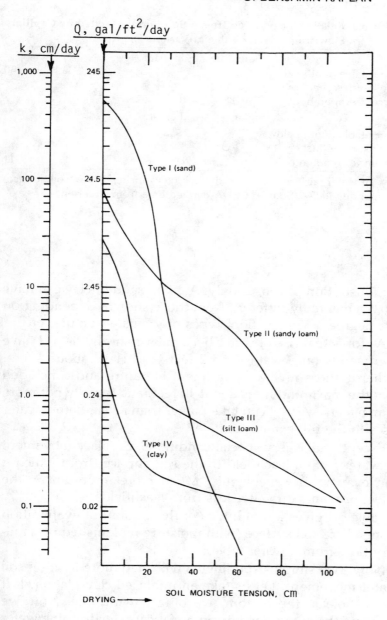

Figure 5.10 Hydraulic conductivity k as a function of soil moisture tension. (Adapted from Otis et al.)[1]

Table 5.1 Relative Evaporation from a Soil Surface Wetted by Capillary Flow from a Water Table at Various Depths[a]

Soil: fine sandy loam						
Water table depth below soil surface, inches	4	16	28	38	43	50
Percent evaporation		88	80	62	33	8	7

Soil: riverbed sand						
Water table depth below soil surface, inches		3	6		10.5		24
Percent evaporation		69	64		58		11

[a]A free water surface on top of the soil (zero inches below soil surface) evaporates 100%. (Source: USDA-derived data from Israelsen.)[3]

for illustration purposes only. A given sand or clay can have curves markedly different from those shown, depending on the degree of compaction, types of grains, structure, etc.

As for actual rise of water by capillarity in the field, I have seen sands raise water 1 to 2 feet, and clays about 4 feet. Other authors reviewed pertinent literature and concluded that the maximum lift is about 10 feet, and usually not more than 5 feet.[2,3] And it can take longer than a year before water rises to the full extent possible.[2]

Except under very special conditions, capillary movement of water from a leachfield to the soil surface (from which it can evaporate) is negligible. Much more important is the effect of plants on water loss from leachfields.

Table 5.1 gives a rough idea of the amount of evaporation from a bare soil surface (with moisture replenished by capillary flow from a water table).

Now, two simple experiments will shed more light on soil water movement. Let us take an unglazed clay planter. If it has a hole in the bottom, we plug it with wax. Now we suspend the pot under a dripping faucet. The dripping water will be absorbed by the dry clay bottom and will start climbing and saturating the sides of the pot (capillary rise). While it is climbing, water does not drip out of the pot, because it is under a tension gradient (created by the capillary pull from the microscopic pores in the pot's clay sides). After the water has climbed to the top of the sides and has saturated all of the microscopic pores, water accumulates over the pot's bottom and drips from the underside. This drip flow will depend on the head or height, h, of water over the bottom, on the thickness of the clay bottom, X, and on the k of the clay material.

The same thing happens when the discharge from a leachfield percolates down through sand and hits a dry soil clay layer which overlies dry sand. Dry sand has very low hydraulic conductivity. (See Figure 5.10.) The dry sand acts like the air below the bottom of the clay pot. As droplets of water form and hang from the underside of the pot's bottom and come together, grow, and drip from the lowest point or protrusion, so does water drip out of the soil clay; once it fills up a pore in the sand below, most of the flow is channeled into this pore, in a sudden discharge.

The last experiment is as follows. Let us pour water into an inclined gutter, and observe two things: One, the surface of the running water is parallel to the bottom of the gutter. Two, if we introduce water into the gutter at a constant rate, steepening the gutter's angle fastens the flow through the gutter and decreases the flow's vertical cross sections.

What happens to flow in the gutter also happens to flow in unconfined aquifers, as shown in Figure 5.11.

An aquifer is any water-bearing soil strata. A confined aquifer is confined by impermeable layers, which act as the walls of a pipe and keep the water under some pressure. In Figure 5.11, if the strata on which trees are growing were impermeable clay, the wet area as drawn (dotted) would be confined.

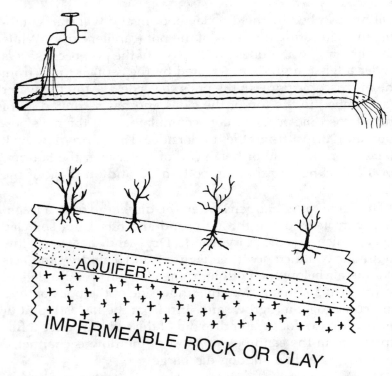

Figure 5.11 Analogy between a gutter and an unconfined aquifer.

5.3 PLANT TRANSPIRATION

Plant roots grow toward moisture, and absorb water (and dissolved nutrients). But usually most of the absorbed water moves to the roots by capillary flow. Roots may help "pump" the water up to the rest of the plant by increasing the osmotic pressure of the absorbed water, or they may just allow the water tension (developed by evaporation from the leaves) to pull the water (and nutrients) up. About 98% of the absorbed water evaporates from the leaves' pores (stomata); this process is called transpiration. Roughly 2 to 4% of the absorbed water is retained in the plant.

About 7 feet of water per year may evaporate from a free

water surface in some desert areas of San Bernardino County. Pastures or lawns transpire about 60 to 80% as much as a free water surface. Small to medium sized trees transpire about 10 to 30 gallons per day. A large eucalyptus may transpire 1 ton of water (290 gallons) per day.

In summary, plants may play a role in the design of septic systems. But one must prevent the roots from plugging drainage or leachline pipes.

5.4 "REST AND DIGEST" SUGGESTION

As of now, practically all the math and physical principles necessary to understand septic systems have been presented. These principles can be a powerful analytical tool when reading pertinent literature or evaluating percolation reports. After you feel you have grasped the principles, we can start applying them and solving problems to increase our understanding even further. Solutions to some of the problems are in Appendix C. Professionals who are in the business of doing percolation test reports (and septic system design), should first try to solve the problems themselves, "sleep over them" if necessary, and only then check the answers in the Appendix.

5.5 PROBLEMS

The truths or principles best remembered are those one discovers oneself. The problems given in this book are designed to help you in this process of discovery. Some of the problems might be tough: they are errors I found in the literature or in procedures employed by experienced engineers. So, don't feel bad if you can't solve them without help. Just try.

5.5.1 Slopes and groundwater

In some communities which have shallow soils (10 feet thick or less) over impermeable bedrock, the sewage discharged through leachlines located at higher elevations may raise the groundwater level and flood leachlines at lower elevations. By means of the Darcy equation, we can estimate how many septic systems can be safely placed at higher or lower elevations.

Let us tackle and solve a tough problem first, so that the following ones will appear easy.

The Unwanted Irrigation Canal

Please refer to Figure 5.12. The story is as follows. An irrigation canal has been recently extended near a house in the desert. It transports water from west to east, near the foot of a mountain chain. The canal is lined beyond the western horizon shown in the figure, but it is unlined from at least one mile east to one mile west of the house.

The owner of the house is worried that the seepage from this unlined canal will raise the groundwater level and flood his leachline. The bottom of his leachline is 5 feet below ground. Does he have a reason to worry? We observe the setting, and see or are informed that:

a. The site is in an alluvial deposits valley. On the north the valley is bounded by a mountain chain of impermeable igneous rock.
b. Throughout the valley, the soil profile is, top to bottom: 0 to 11 feet, sand; 11 to 260 feet, solid impermeable clay.
c. The valley terrain and the soil strata slope at a rate of 1 foot down per thousand feet horizontal distance west to east. The canal slope is identical. The terrain and soil strata slope at a rate of 20 feet down per thousand feet horizontal distance north to south.
d. The 11-foot-thick sand stratum over the clay is perfectly

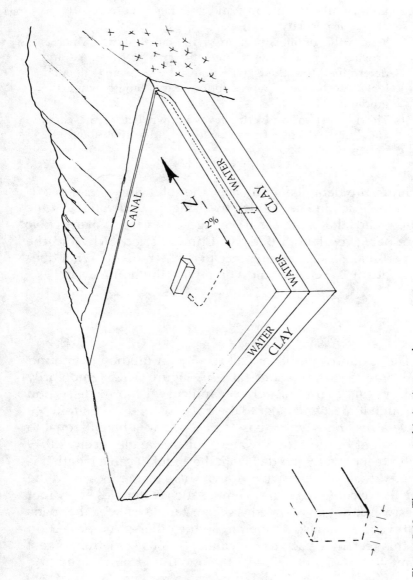

Figure 5.12   The unwanted irrigation canal.

uniform in all directions, and its k has been measured to be 1 foot per day.

e. Where the unlined portion of the canal starts, water is flowing. An expert examined this area of the canal and determined that the canal will lose by seepage 0.19 cubic feet of water per day per foot of canal length when the gradient is 2%.

f. The highest groundwater level known before the canal was built is 10 feet from the surface (one foot above the impermeable clay).

From the data above, answer the following questions: 1. Will the leachline be flooded by rising groundwater? 2. One often reads that earth canal seepage losses are so many cubic feet per given length per given time. Period. Why did the expert make a special reference to a 2% gradient? What does this mean? 3. What would you advise the homeowner?

Answers

The groundwater level will rise by an unknown amount. We draw a groundwater table in Figure 5.12, temporarily half way up to the soil surface, and cut a 1-foot-wide section through it, perpendicular to the direction of flow. So now we have a rectangle with a base of 1 foot and a height equal to the height of the groundwater. This rectangle receives 0.19 cubic feet of water per day from the 1 foot of canal length it is related to. We could have drawn this rectangle close to the canal. We could have drawn more such rectangles every foot westward. But all we need is one rectangle, as what happens in this one will also happen in all the others. The projection of this rectangle back to the canal gives us something like a giant "tube" of wet soil, like the one used to measure k. As shown in Figure 5.12, this tube has a rectangular cross section; its base is 1 foot wide and its height is the unknown height of the groundwater. It slopes at a gradient of 0.02, or 2%. Therefore, using the modified Darcy formula, we have:

$$Q/S = k \sin\Theta$$

$$0.19 \text{ (ft}^3 \text{ per day)}/S \text{ (ft}^2) = 1 \text{ (ft per day)} \times 0.02$$

Solving for S, S = 9.5 square feet. S is the cross-sectional area of the tube. The base times the height is 1 (ft) x ?(ft) = 9.5 square feet. Hence, the height of the groundwater will be 9.5 feet. The leachline will be flooded.

From Darcy's formula we know that we can neglect the fact that another gradient exists—1 per 1,000 (or 0.1% or 0.001). Since the flow per unit area is directly proportional to the sine of Θ when the k's are the same, the flow on the 2% gradient is about 20 times larger than that on the 0.1% gradient. And how do we know the value of the sines? Well, the sine of a gradient is numerically the same as the gradient, from 0 to about 0.20. That is, the sine of the angle which corresponds to a slope of 1% or 0.01 is 0.01; to a slope of 0.05, is 0.05; to a slope of 0.20, is 0.20.

The expert specified a seepage under a gradient of 2% because he knew that the rate of infiltration depends on the gradient. When infiltration just starts, the gradient is huge. When the water starts moving from the canal surface to the clay layer, the gradient is more than 1. (h + X)/X is more than 1; h is the pressure of the water on the canal surface, and X is the distance traveled by water through the soil. When the water is moving along the 2% slope for 100 or 1,000 feet, the effect of h is negligible, so that H = 2 feet when X = 100 feet, and 20 feet when X = 1,000 feet; and so the gradient is only 0.02 or 2%. Note that H is not just the height of the water from the canal down to the clay, or 11 feet. H is the difference in water level elevation from the canal to any point. Published seepage data usually refer to canals in use for many years, so that X is large and the seepage rate is stable.

The homeowner has options. He can go to the canal owners or designers and demand that it be lined, on the basis of the calculations shown above. Or, he can go to the nearest

Agricultural Extension office and find which trees suited to the site conditions can be planted to lower the groundwater level.The amount of seepage is quite modest: 0.19 cubic feet per day per lineal foot of canal is a mere 1.5 gallons per day per lineal foot of canal, or about 30 gallons per day per each 20 feet in an east-to-west direction.

Perennial fruit trees or ornamentals may do the trick. Eucalyptus trees have been used to drain swamps. Tamarisks do too well in the desert, near groundwater tables. They do lower groundwater very effectively, but are now considered to be "pests," almost impossible to eradicate once they take hold.

Now let us change some parameters, and let the reader do the figuring. How high would the water rise if the slope were 10% instead of 2%?

If the slope is 2%, the leachline is 500 feet south of the canal, and the clay layer is rather permeable, with a k of 0.01 foot/day, would the leachline be affected?

Upslope Development

Tract houses have been built over 15 feet of decomposed granite soil overlaying undecomposed impermeable granite (Figure 5.13). The slope of the soil and of the granite rock surface is 10% downward west to east. The soil is uniform and has a k of 1 foot/day. The tract is square, 1,000 x 1,000 feet. The lots measure 100 x 100 feet, and each lot has a 3-bedroom house, a 1,000-gallon septic tank, a 90-foot leachline parallel to contour lines (perpendicular to the slope), and an average discharge of 300 gallons per day. There is no groundwater. Disregard the effect of rainfall or lawn irrigation or trees when you answer this question: Would you buy a house in this tract if you see there is one house upslope? If you see three houses upslope? Would you buy a house at all?

Let us estimate the effect of one house on creating a water table. First we convert everything to the same units. 300 gallons per day is about 40 cubic feet per day. The 40 cubic

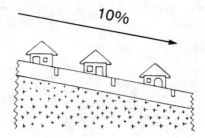

Figure 5.13 Cross section of upslope development.

feet are discharged over 90 to 100 feet, say 0.40 cubic feet per foot of property line perpendicular to slope. Then,

$$0.40 \ (ft^3/day)/S(ft^2) = 1 \ (ft/day) \ sin\Theta$$

Since $sin\Theta = 0.1$, $S = 4$ square feet. The base of the rectangle of area S is 1 foot. So, the rise in groundwater level is 4 feet. Three houses upslope would raise it 12 feet. I would not buy any house. About 70% of the houses may have severe sewage disposal problems.

5.5.2 Measuring k Correctly

It is routine for soil engineers to determine k after increasing h to about 30 feet (usually with gas pressure). This procedure increases the Q, and is quite convenient when one tests clays that barely drip unless they are subjected to a high h. The method can be found in Army Manual EM 1110-2-1906, and in California Administrative Code Title 23, Chapter 3, Subchapter 15, Appendix I. It was designed to test the permeability of clays that are to be used as liners to contain hazardous waste.

However, on occasion a consultant will determine a k for a leachfield design problem by the procedure described in the

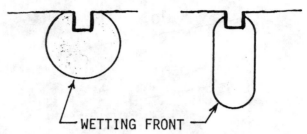

Figure 5.14 Wetting fronts.

paragraph above, and his or her answer will be wrong. We'll see why later in Appendix C.

An EPA manual[4] has presented a formula for determining the average k of horizontally stratified soils (for instance, a layer of sand over a layer of clay over a layer of sand). It is

$$k = \frac{d_1 + d_2 + \ldots d_n}{\frac{d_1}{k_1} + \frac{d_2}{k_2} + \ldots \frac{d_n}{k_n}}$$

where k is the harmonic mean k, and d_n and k_n are the thickness and the k of stratum number n. This formula also provides the unwary consultant with opportunities to make mistakes when dealing with septic systems. The reason for the potential mistakes is not difficult to figure out, if someone has "digested" this chapter and takes time to solve this riddle, or turns to Appendix C, where the solution is revealed. (Hint: picture vertical water movement in various kinds of stratified soils, and deduce the effect of various h values on measured k values.)

5.5.3 Wetting Fronts

Someone introduces about 5 gallons of water into holes dug into two types of soil: one is a clay, the other is a sand. After all the water is absorbed, cross sections of the soils look as shown in Figure 5.14. Which is the clay soil and which is the sand? Why? The answer is in Appendix C.

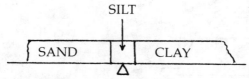

Figure 5.15 Balanced weights of dry sand and dry clay.

5.5.4 Evaporation

A leachline's bottom is about 38 inches below the soil surface. The soil is a fine sandy loam. The leachline is 100 feet long and 3 feet wide, and receives 300 gallons of sewage per day. Assume that the sewage effluent acts like a water table near the bottom of the leachline and that evaporation is as shown in Table 5.1. The evaporation from a free water surface is 4 feet per year.

The sewage not evaporated percolates downward. What percentage of the sewage discharged is lost to evaporation? The answer is in Appendix C.

5.5.5 The Balance

Dry blocks of sand, of silt, and of clay are in a balance, as shown in Figure 5.15. The silt block is taken out, moistened to near saturation, and placed back between the clay and the sand, exactly in the middle and without losing its shape or displacing the sand or the clay. There is good contact between the silt and sand, and silt and clay surfaces. Which way will the balance tilt after a few minutes? Why? The answer is in Appendix C.

5.5.6 Slope Seepage

Figure 5.16 shows a new house (on terrace deposits) up above a valley in a desert climate (no rain). A leachline between the house and the slope discharges into a stratum of sand. The leachline is heavily loaded with 5 or more gallons of sewage per square foot of leachline bottom per day. The

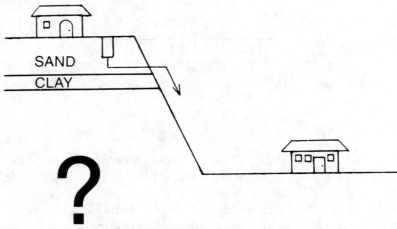

Figure 5.16   Slope seepage.

sewage may move down, hit the clay, move laterally, and seep out or surface as indicated with an arrow. Which condition will make this seepage more likely: (a) clay overlying coarse sand, or (b) clay overlying more clay? The answer can be found in Appendix C.

REFERENCES

1. Otis, R., et al. 1977. Alternatives for Small Wastewater Treatment Systems. EPA Technology Transfer Pub. No. EPA-625/4-77-011.
2. Tolman, C.F. 1937. *Ground Water*. McGraw-Hill, New York.
3. Israelsen, O. 1950. *Irrigation Principles and Practices* (2nd edition). John Wiley & Sons, New York.
4. U.S. EPA. 1981. Process Design Manual for Land Treatment of Municipal Wastewater. EPA-625/1-81-013.

6

The Percolation Test, or "Perk" Test

A leachline (or seepage pit) of a given size is likely to last longer if installed in a sandy rather than in a clayey soil. So, people are prone to feel that a soil's texture determines its permeability, and hence its suitability for leachfields. But we should remember that permeability is influenced by the extent of clogging mat development, and that texture and permeability are generally weakly correlated (Chapter 4). Occasionally, texture-permeability correlations may be found in published literature, but only as a local phenomenon that cannot be duplicated elsewhere, particularly if soils are high in clay.[1]

Table 6.1 compares textures and relative permeabilities of different soils. Its figures were taken from data reported by Hantzsche et al.[2] Perhaps this poor correlation was known to exist early in this century; perhaps not. Either way, since

Table 6.1 Relative Permeability of Various Soils as Measured in Minutes per Inch Through Percolation Tests[a]

Texture	Mean mpi	Expected range: mean plus minus one standard deviation
Loamy sand	32	< 5 to 78
Sandy loam	17	< 5 to 28
Loam	25	< 5 to 73
Clay loam	33	< 5 to 97
Clay	152	63 to over 240

[a]The soils are ranked texturally. High mpi means low permeability. Each type of soil is represented by at least 10 measurements.

Ryon's time, relative permeability has been used to rate soil suitability for septic systems.

6.1 HISTORICAL BACKGROUND AND DEVELOPMENT OF THE PERK TEST

In the 1920s a New York engineer, Henry Ryon, tried to estimate the permeability of soils near "failing" and nonfailing leachfields by means of what came to be called a percolation test, or perk test. According to Winneberger,[3] Ryon dug a one-foot-square hole down to leachfield depth, wetted it, refilled it, and measured how long it took for the water level to go down one inch from an initial surface head of 6 inches. Then, he associated this crude measure of permeability to leachfield size and longevity, and developed soil-absorption-area standards to (try to) achieve a minimum leachfield life-span of 20 years before "failure."

The percolation test does not measure percolation: we can't determine how fast water moves through a soil by measuring how fast it disappears from a hole in the soil. It doesn't measure infiltration either; infiltration measurements require a flat surface and other conditions which are not part of the perk test procedure. Since it measures nothing but what it happens to measure, i.e., a gross relative permeability, I prefer to use the honest term "perk," or "perking," instead of "percolation testing."

The U.S. Health, Education, and Welfare Department tried to refine Ryon's methodology in 1949, and commissioned a series of studies in this regard. The studies' conclusions sometimes were patently flawed, as we will see. Yet they influenced the standard perk test procedure described in 1957 by the U.S. Public Health Service[5] and in 1980 by the U.S. EPA.[6]

Weibel et al.[4] conducted the HEW-commissioned studies. In order to improve Ryon's crude methodology, they suggested: "(1) Use a smaller bored hole in place of the 1-foot-square dug hole; (2) use a more specific procedure to insure saturation, i.e., soak the soil overnight before making test; (3) use the percolation rate over the later period of the test, rather than the average rate, as being the more nearly true rate; (4) use several replicate tests in any area under consideration." Weibel et al. conducted experiments to explore and define some of these suggestions. The outcome of their findings follows.

6.1.1 Size of Bored Hole

Weibel et al. tested three types of holes in three types of soils.[4] The hole types were 4-inch-diameter, 8-inch-diameter, and 12 inches by 12 inches square. They obtained means from three test holes of the same size, in each of the three types of soil. They reported: "there appears to be no overall correlation between percolation rate and size of hole," and "it appears that the differences in rates between replicate tests within the same soils are greater than the differences in average rates between different sized holes in the same lots (soils)".[4] In plain English, they said that the experimental error was so large (mainly because of the small number of replicates) that no differences due to hole sizes were noticed. They also quoted literature regarding statistical studies which "indicated that from 6 to as many as 22 individual infiltration tests are necessary to estimate closely the representative rate for a specific soil site . . . with relatively homogeneous soils." Yet they averaged only 3 tests per mean, not 6 or 22.

Curiously, the committee that prepared the USPH Manual[5] read the sentences quoted above, yet concluded that the type of hole makes no difference as long as it is "4 to 12 inches in horizontal dimension." Now, many kids who have played by the seashore and filled holes in the sand with water know (and may still remember) that water disappears much faster from a smaller-diameter hole. I guess there were no such kids on the committee.

Even more curiously, this error was not detected for more than a decade. Unchallenged, it was carried over in respectable engineering textbooks. And just last month an engineer from another county told me, while brandishing his copy of the USPHS Manual,[5] "here it says that it (hole diameter) doesn't make a difference."

Winneberger, who knew better, stated in his book[3] that, when he was a consultant and had to produce satisfactory perk results on low-permeability soils, he used 4-inch-diameter holes.

Olivieri and Roche hypothesized in 1979 that a perk rate obtained from a test hole of a given diameter could be multiplied by a specific correction factor to yield a rate specific for a hole of different diameter.[7] The correction factor is equal to the ratio volume/surface of one hole, divided by the volume/surface of the other hole. This hypothesis has merit, and it ahs been endorsed enthusiastically in some quarters.[8] But it cannot be justified as a panacea on theoretical or practical grounds, as will be seen later.

Finally, in 1980 the EPA Manual[6] specified a 6-inch hole diameter in its standard perk test procedure (but referred to 6- to 9-inch diam. holes elsewhere). And in 1981 Van Kirk et al.[8] promoted the adoption of the 6-inch-diameter hole by Caltrans (California Department of Transportation).

The 6-inch-diameter standard is now in wide use, and the EPA sewage-disposal tables are "geared to work" with this diameter. For these two main reasons, this is to be our own standard hole diameter.

6.1.2 Soaking the Hole Overnight

(Here we have another opportunity to apply what we learned in the previous chapter regarding X and h.).

Weibel et al.[4] recommended that, before testing, the hole be filled with 14 inches of water, and be kept full to that level

. . . for at least 4 hours and preferably overnight (Important: not that the word "preferably" may result in capriciousness).

Allow the soil to swell overnight. This saturation procedure insures that the soil is given the opportunity to swell and approach the condition that it will be in during the wettest season of the year. Thus, the test will give comparable results in the same soil whether made in a dry or a wet season.

(a) Floating Indicator

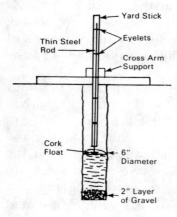

(b) Fixed Indicator

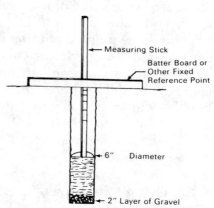

Figure 6.1 Standard percolation test hole. (Adapted from Otis et al.[6]) (a) Precise measurement of surface head with the help of a float. (b) Less precise measurement of surface head with a stick (or steel tape).

. . . In sandy soils containing little or no clay, the swelling procedure is not essential . . .

True enough, some soils contain expanding clays of the smectite family (montmorillonite, bentonite, nontronite). Such clays' particles immobilize layer upon layer of water molecules around them (20 to more than 80), and push away from each other; i.e., they "swell" and restrict water flow through the soil pores, and lower the hydraulic conductivity k. But, what about the distance to the wetting front, X? Weibel et al. forgot that the gradient changes with X, among other things. If some perk testers prefer to saturate overnight and others do not, their "X's" and their results are likely to be different.

Figure 6.2(a) shows how the perk rate in inches per hour changes through time as water in the test hole is absorbed and replenished. The rate seems to "stabilize" in the first run at about 0.9 inches/hour, but after an overnight soaking, the rate "stabilizes" at about 0.5 inches/hour. However, if the measurement continues for 120 days, we would see as in Figure 6.2(b) that the perk rate keeps on decreasing: presumably, as the "average" X increases, the "average" gradient approaches the value of unity, and the perk rate approaches a constant value similar to k.

It should be pointed out that things are actually a bit more complicated. The perk rate curves may be affected not only by swelling of clays and increases in X, but also by soil-aggregate stability (pores may collapse), by entrapped air bubbles within the soil pores, by microbial growth and activity, and by uniformity of soil. Occasionally one might see that a perk rate increases after a day or two, before it starts decreasing again.

6.1.3 Using "Stabilized" Instead of Average Perk Rates

If one tries to measure a parachutist's falling speed, it doesn't make any sense to measure his speeds from the

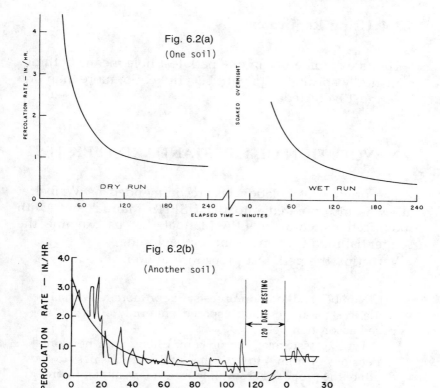

Figure 6.2   Trend of percolation rates in two different field test holes with time. (Source: Weibel et al.[4])

moment he jumps from the plane to the moment he lands, and average all the measurements. The "average speed" would depend on how long it took for the parachute to open. And one would be adding and averaging accelerated, variable free-fall speeds together with a (terminal) constant (opened-parachute) speed—in other words, adding apples and oranges. In the same vein, one has to measure the "terminal" or "stabilized" perk rate. But no one can wait 120 days or so for a true stabilized rate; so, whatever is measured after a conveniently long period of about one day is called "stabilized."

6.1.4 Using Replicates

The rate obtained from one perk test hole means nothing, statistically speaking. It might be a fluke. So, more than one hole must be tested.

6.2 EVOLUTION OF THE STANDARD PERK TEST

The original percolation test recommended by Weibel et al.[4] was incorporated into the USPHS Manual and, with modifications, into the EPA Manual. Let us examine the sequential modifications from the beginning.

Verbatim, the perk test procedure of Weibel et al.[4] was:

1. Number and location of tests. Six or more tests shall be made in separate test holes spaced uniformly over the proposed absorption field site.

2. Type of test hole. Dig or bore a hole, with horizontal dimensions of from 4 to 12 inches and vertical sides, to the depth of the proposed absorption trench (leachline). In order to save time, labor, and volume of water required per test, the holes can be bored with a 4-inch auger.

3. Preparation of test hole. Carefully scratch the bottom and sides of the hole with a knife blade or sharp pointed instrument, in order to remove any smeared soil surfaces and to provide a natural soil interface into which water may percolate. Remove all loose materials from the hole. Add 2 inches of coarse sand or fine gravel to protect the bottom from scouring and sediment.

4. Saturation and swelling of the soil. Carefully fill the hole with clear water to a minimum of 12 inches over the gravel. By refilling if necessary, or by supplying a surplus reservoir of water, such as in an automatic siphon, keep water in the hole for at least 4 hours and preferably overnight. Allow the soil to swell overnight. This saturation procedure insures that the soil is given the opportunity to swell and approach the condition that it will be in during the wettest season of the year.

Thus, the test will give comparable results in the same soil whether made in a dry or a wet season.

In sandy soils containing little or no clay, the swelling procedure is not essential and the test may be made as described under item 5C, after the water from one filling of the hole has completely seeped away.

5. Percolation rate measurement. With the exception of sandy soils, percolation measurements shall be made on the day following the procedure described under item 4, above.

A. If water remains in the test hole after the overnight swelling period, adjust the depth to approximately 6 inches over the gravel. From a fixed reference point, measure the drop in water level over a 30-minute period. This drop is used to calculate the percolation rate.

B. If no water remains in the hole after the overnight swelling period, add clear water to bring the depth of water to approximately 6 inches over the gravel. From a fixed reference point, measure the drop in water level at approximately 30-minute intervals for 4 hours, refilling 6 inches over the gravel as necessary. The drop that occurs during the final 30-minute period is used to calculate the percolation rate. [Note: one irrelevant sentence is omitted.]

C. In sandy soils (or other soils in which the first 6 inches of water seeps away in less than 30 minutes, after the overnight swelling period) the time interval between measurements shall be taken as 10 minutes and the test run for one hour. The drop that occurs during the final 10 minutes is used to calculate the percolation rate.

The USPHS procedure[5] is practically identical. The main difference is that, regarding point 4 above, the USPHS text explains that saturation refers to filling of soil pores and that swelling refers to swelling of the clays.

The EPA procedure[6] contains the following changes to Weibel's points 1 through 5 above:

1. Minimum 3 tests; more if soil conditions are highly variable.
2. The diameter of the test hole is 6 inches; the hole is dug or

bored to the proposed (leachline) depth or to the most limiting soil horizon.

3. (No change.)

4. In sandy soils with little or no clay, if the 12 inches of water seep completely away in a 10-minute period, and seep away again (after refilling) in the next 10 minute period, the test can proceed immediately.

5. Except in the case immediately above, perk rates are measured not less than 15 hours nor more than 30 hours after the beginning of the saturation/swelling period. The water level is not allowed to rise more than 6 inches over the 2 inches of gravel. Measurements are precise to the nearest $1/16$ of an inch.

   A. At least 3 measurements are made. Measurements are continued every 30 minutes until two in succession do not vary by more than $1/16$ of an inch.

   B. Same as A above.

   C. (No change.)

6. The EPA standard method has an additional section entitled "Calculation of the percolation rate." Per this section, one divides the time interval used between measurements by the magnitude of the last water level drop, and obtains a percolation rate in minutes per inch. Then, the rates for each hole are averaged; but if the rates vary by more than 20 minutes per inch, they are not averaged because they reflect different soils within a site.

6.3 EVALUATION OF AND IMPROVEMENTS NEEDED IN THE EPA STANDARD PERK METHODOLOGY

The EPA Manual[5] authors improved perk methodology in various ways. They

a. reduced the minimum number of tests to a more reasonable figure — 3 instead of 6

b. standardized the diameter of the test hole at 6 inches (variable diameters may result in variable measurements)

c. defined more precisely when measurements were to be

made (more than 15 but less than 30 hours after the first wetting of the hole—this also decreases variability of results)

d. specified that the water level must not exceed 8 inches over the hole bottom during testing (higher levels increase h and sometimes may cause something similar to backflow of water into the hole if the level goes down rapidly)

e. specified that minimum precision of measurements is $1/16$ of an inch (this may remind perk testers that sloppy readings may invalidate conclusions as to whether a lot is or is not suitable, in borderline cases near the conventional maximum perk of 60 minutes per inch)

In spite of all these improvements, some deficiencies remain in the EPA perk procedures:

a. In some soils, the abundance of cobbles makes it impossible to use a 6-inch-diameter hole. Larger diameters are unavoidable.

b. A standard cannot be capricious. Weibel's soaking procedure, uncorrected, is still found in the EPA procedure. On a given soil, if one perk tester prefers to maintain at least 12 inches of water overnight during the swelling/saturation (or soaking) period, while another one prefers not to, each will saturate the soil to a different extent, the gradients will differ, and so will the results. The same is true if one tester maintains the minimum 12 inches above the bottom of the hole while another tester maintains 60 inches.

c. It is not enough to remove a hole's smeared surfaces. Often, when digging or drilling with a power auger, the soil is severely compacted behind the smeared surfaces, and has to be "peeled." Van Kirk et al.[8] reported that in one soil, power-auger-drilled holes had a mean perk of 118.5 minutes per inch, while hand-augered holes had a mean perk of 3.1 minutes per inch; in another soil the respective means were 33.8 and 0.8 minutes per inch.

d. The EPA procedure calls for removing soil that sloughs down into the test hole during the soaking period. This removal often ends up compacting hole sidewalls, plugging soil pores, changing the hole's diameter, and leaving a real mess.

e. The water level in a test hole falls only about half an inch over a 30-minute measuring interval if the soil has a 60-minutes-per-inch perk time. So, the surface head h changes little. But if the soil has a perk time of 1 minute per inch, the hole might dry out between the 10-minute measurement intervals, or the rate of fall in h may slow down as h decreases; so, the measurements may mean nothing.

f. A conceptual error in the EPA procedure is that it expresses the measurements as a rate in minutes per inch (rather than in inches per minute or per hour, as Weibel et al. did[4]). Furthermore, it calls for averaging such "rates" (actually reciprocals of true rates) from 3 or more holes (which do not differ by more than 20 minutes per inch).

Perk rates are expressed in units of space per time, say inches per minute, or ipm. Perk times may be expressed in minutes per inch, or mpi. One cannot calculate the average of some mpi values, divide 1 by the result, and obtain a real average rate, ipm. It is mathematically impossible.*

Winneberger[3] noted this error. He analyzed experimental data on percolation rates, and concluded that the best "average" of the data was the geometric mean rather than the arithmetic mean. (This conclusion is unwarranted: some soils yield perk rates which are distributed on a log-normal curve, and the geometric mean can be used to measure central tendency but not to average infiltration; furthermore,

---

*For instance, let us fancy two test holes, one with a perk time of 1 mpi, and the other with a time of 20 mpi. The average mpi is 21/2 = 10.5 mpi. The reciprocal of this is 0.095, but it is not 0.095 ipm. The average ipm is calculated as follows. The first hole's ipm is 1/1 mpi = 1 ipm; the second hole's ipm is 1/20 mpi = 0.05 ipm. The average ipm is therefore (1 + 0.05)/2 = 0.525 ipm, which is quite different from the unreal "average," 0.095 ipm.

some soils yield rates distributed on a normal curve, and the arithmetic mean is the proper parameter.)*

So, Winneberger had three recommendations for avoiding EPA's error: One, use the geometric mean. Two, discard the highest and lowest mpi "rates" before calculating the arithmetic mean. And three, calculate the reciprocals of the mpi, average the results, and then calculate the reciprocal of this average to obtain the average mpi.

The first recommendation is itself in error, as explained below. The second one merely improves a bit the result of an improper mathematical operation. The third one is correct.

The geometric mean is not the proper parameter. To explain this graphically, let us simplify matters and conduct a "mental experiment." Let us "test" the average perk rate or mpi in a trench, half of which runs through a very permeable sand, and half of which runs through a practically impermeable clay. Half of the perk tests yield 2 ipm (sand), and the other half yield 0.005 ipm (clay). To simplify matters further, let us assume that one representative hole is tested in the sand, and one in the clay. (The results will be identical whether we perform calculations with 1 and 1 hole or with 100 and 100 holes.) Intuitively, we know that half the length of the trench will absorb water at a rate proportional to 2 ipm, and the other half will absorb practically no water. So, the average water absorption over the whole trench will be related to the average of "two plus (nearly) zero" ipm, which is 1 ipm.

Now let's see how accurate are the arithmetic mean (a.m.) and the geometric mean (g.m.) and their reciprocals (1/a.m. and 1/g.m.).

means of:	a.m.	1/a.m.	g.m.	1/g.m.
2 and 0.005 ipm	1	1	0.1	10
0.5 and 200 mpi	100	0.01	10	0.1

---

*The arithmetic mean is "the average" we all know about. To obtain the geometric mean of n numbers, say a, b, c . . . n, we multiply these numbers and take the nth root of the product. For instance, the geometric mean of 1 and 9 is the square root of 1×9. The geometric mean of 1, 9, and 10 is the cubic root of $1 \times 9 \times 10$.

It is obvious that only the a.m. of the true rate, inches per minute, and its reciprocal (minutes per inch), do not contradict what we know to be accurate.

As for the other deficiencies in the EPA methodology, a, b, c, d, and e mentioned above, they are amenable to corrections. Respectively:

a. Large holes can be dug, and the rates measured can be corrected in the most conservative direction by means of the Oliveri-Roche correction factors. (See sections 6.1.1 and 6.4.8.)

b. The "capricious" variability in surface head and time of soaking can be mitigated by requiring that a given volume of water be absorbed. For instance, one can invert a 5-gallon bottle full of water over the test hole, 12 inches above the hole bottom. Then one can start measurements when the bottle is empty (in very sandy soils, with perk times of 2 to 4 mpi, the bottle empties in about 15 to 30 minutes), or on the following day (in clayey soils with a perk of 120 mpi, the bottle will be about $1/4$ full the next day; and with 400 mpi, it will be about $1/2$ full).

c. One can bore a 4-inch diameter hole and then "peel off" the compacted sidewalls until the diameter is 6 inches. This is done by inserting a nail (or the tip of a pocket knife) into the sidewall and moving it centripetally to peel off sliver after sliver of sidewall.

d. One can support the sidewalls to minimize the sloughing off during the soaking or testing period. A common way is to insert an 8-inch long piece of 2- to 4-inch diameter perforated pipe into the test hole, and fill with pea gravel the space between the pipe and the hole sidewall. This is called "gravel packing."

e. One can specify that the measurement intervals must be shortened so that h does not fall more than 3 inches. (I have observed that, in very sandy soils, measurements do not vary much if the fall in h is restricted to a maximum of 3 inches.)

Corrections have been incorporated into a streamlined version of the EPA's standard perk procedures, presented in Appendix I.

And now, the moment I have been waiting for: Let's go to the next section and have some fun with practical and theoretical problems.

6.4 PROBLEMS

Let us start by solving the kind of practical problems that face perk testers and environmental health professionals. Then we'll entertain ourselves with more esoteric problems.

6.4.1 Gravel Packing of Perk Test Holes

Perk testers who have had the patience to get to this point will now be rewarded. During the years I have been reviewing perk reports, I have noticed that perk testers invariably do laborious, time-consuming calculations to correct for the effect of gravel packing on measured perk rates. The formula derived below will make such calculations short and easy. (If you do perk testing, hopefully you'll be able to save more from one use of this formula than what you spent for this book.)

Figure 6.3 shows a perk test hole with radius r_2 and, inside it, a perforated pipe with radius r_1. Loose gravel is placed just

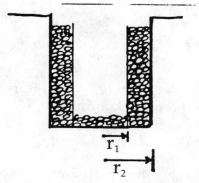

Figure 6.3 Vertical cross section through a gravel-packed perk test hole.

outside the pipe to fill the cylindrical space between r_2 and r_1. The gravel occupies a given volume, and displaces that same volume of water out of the hole. Since there is less water within the hole, the water level falls faster than if there is no gravel.

The volume of a cylinder is $V = \pi r^2 h$, where h is its height. Let us fill the hole with water. The total volume of water (Vw) in the hole with no gravel is

$$Vw = \pi r_2^2 h;$$

and the volume of water within the pipe is $Vp = \pi r_1^2 h$. The volume of water between the pipe and the hole sidewall is

$$Vw'' = (\pi r_2^2 h - \pi r_1^2 h).$$

We have that $Vw = Vp + Vw''$. The volume of water not displaced by the gravel will be equal to the gravel voids or pore space. And if the porosity of the gravel is P, the total water volume in the hole is

$$Vw' = Vp + PVw''.$$

The "gravel correction factor" is equal to the ratio of water volumes with and without the gravel: Vw'/Vw.

We have that

$$Vw'' = \pi r_2^2 h - \pi r_1^2 h; \text{ let } r_2/r_1 = C. \text{ Then,}$$

$$Vw'' = \pi h r_1^2 [C^2 - 1] \quad \text{and} \quad Vw = \pi C^2 r_1^2 h$$

So $Vw'/Vw = \{[\pi r_1^2 h] + P\pi r_1^2 h (C^2 - 1)\}/\{\pi C^2 r_1^2 h\}$ and after simplifying, we get a cookbook formula:

$$\mathbf{Vw'/Vw = [1 + P (C^2 - 1)]/C^2}$$

As an exercise, let us assume that a gravel-packed hole is 6 inches in diameter, the pipe is 2 inches in diameter, the gravel porosity is 33%, or 0.33, and the stabilized rate is 0.5 ipm (equivalent to 2 mpi). What is the true rate (no gravel)? The correction factor is

$$\{1 + 0.33 \, [(6/2)^2 - 1]\}/(6/2)^2 = 0.405$$

Hence, the true rate is

$$0.5 \times 0.405 = 0.2 \text{ ipm (5 mpi)}$$

Another problem is to determine P, the porosity or voids in the gravel. When doing this, soil or civil engineers follow ASTM Standard C 29–78, and commit two types of errors.

When one drops gravel into a test hole, the gravel remains loose and uncompacted, as it should (tamping could compact the hole sidewall). The hole can't be shaken to settle or compact the gravel. But the ASTM standard calls for tamping with a rod, and the measuring bucket can be shaken. So, the ASTM method tends to underestimate the actual value of P.

The other type of error arises when the engineer uses the familiar silica sand (used to measure soil compaction) instead of gravel. In addition to the compaction error described above, sand retains water in its small pores. So, although the sand might have an ASTM P of 38%, the usable P might be at most 35%.

The easiest way to determine P in the field is: Take two identical tin cans (about 23 oz.), A and B. Fill A with loose gravel. Do not shake. Fill B with water. Measure the height of water h. Pour water from B into A until full to the rim. Collect any spilled water in a tray and return to B. Measure the new water level h' in B. Now, P = (h – h')/h.

(Appendix D has a table with correction factors for various values of C and of P.)

6.4.2 Accuracy of mpi or ipm Meausurement

Some perk testers drill test holes down to 5 feet below ground, and lower the tip of their measuring tapes or rods down to the water level, about 4.5 feet from the soil surface, and then measure changes in water level. If the measuring tape or rod wanders from the vertical and the measuring tip is 6 inches away from a plumb line, what is the measurement error? What is the measurement error if the respective figures are 1 foot down to the water level and 3 inches off plumb? (The answers are in Appendix D.)

6.4.3 Effect of Precision of Measurements on Accuracy

First let us define precision and accuracy. Take the statement, "my foot is 120.00 inches long." this figure is very precise: it has 5 significant figures (number of digits to the left and right of the period, 3 + 2 = 5). But this figure is very inaccurate, about 108 inches off the mark. Precision refers to sensitivity of measurement (i.e., to the nearest yard, or micron, or angstrom). Accuracy refers to truth.

In a typical jurisdiction, the maximum allowable perk time for use of septic systems is 60 mpi. Assuming that a jursidiction requires a three test minimum, $^1/_{16}$–inch precision, and averaging of mpi, and if the soils are perfectly uniform:

What percentage of soils which actually have perk times of 60 mpi will appear to be "over 60 mpi" due to precision error and chance (and be rejected)? What percentage of soils which actually have perk times just above 60 mpi will appear to be "less than 60 mpi" (and pass)?

Solve this same problem for $^1/_8$–inch precision (Your answers might be as good as mine. See Appendix D for answers.)

6.4.4 **Rate of h Fall in a Perforated Tin Can**

When a test hole is first wetted, during the instant that X is negligible in comparison to h, the h (height of the water column or surface head) should decrease in a manner similar to that seen in a leaky bottom, perforated tin can with a layer of soil covering all perforations.

If the hole's infiltration surfaces are very smeared (or compacted) so that their permeability is a tiny fraction of the permeability of the soil around the hole, h should decrease as if the hole were a leaky tin can suspended in the air, until X grows (if it grows at all) and reduces the gradient.

We fill a tin can with water. The water starts leaking through a hole in the bottom. The water level h goes down. We start measuring the decrease in h when h has gone down to a level of 10.0 inches above the bottom. One second after that, h is at 9.0 inches above the bottom.

a. Derive a formula and calculate the value of h after 2, 4, 8, 16, and 32 seconds. The formula is worked out below.

b. Same as above, but now the can's body is not made out of tin, but of a thin layer of soil or fritted glass with pores throughout. (The answer is in Appendix D.)

The rate of decrease in the height of the water level h is proportional to the pressure on the bottom of the can, which is equal to h. Hence, the rate of decrease is

$$-dh/dt = ch$$

where
c = a proportionality constant
t = time

Another way of expressing this equality is $-d(\ln h) = cdt$. Integrating, we have that

$$\ln h = -ct + c', \text{ or}$$
$$h = e^{c'}e^{-ct}$$

e is the base of natural logarithms; e raised to a constant is a new constant, C, so

$$h = Ce^{-ct}$$

Now, we know that when t = 0, h = 10 inches; therefore (since e raised to the zero power is equal to one), we have that C = 10 inches. And we also know that, when t = 1 second, h = 9 inches. Therefore,

$$9 = 10e^{-c}$$

Then, c = $\ln^{10}/_9$ = ln1.11 = 0.4.

Now that we know the values of C and c, h = $10e^{-0.4t}$ The reader might wish to plug in the values given for t and plot the graph of h versus time.

6.4.5 Effect of Changing h on Perk Rate

Some sentences below have been taken from Winneberger's generally informative and witty book.[3] They do, however, contain errors which the reader might be able to detect.

After the soaking period, after the rate of fall had "stabilized," the perk rate in one test hole was 54 inches/hour with the water level at 6 inches above bottom; but it was 100 inches/hour after the water level was raised to 8 inches above bottom. Winneberger states, "So far as water columns as such were concerned, the hydraulic head was increased 1.33 times. But the rate change, which in theory might have increased proportionately from 54 to 72 in./hr was actually 100 in./hr" (sic).

And in another hole, the water level was decreased from the 8-inch level to the 6-inch level, and the respective rates decreased from 28 to 10 inches/hour—this is not proportionate either. Winneberger's explanation is almost accurate: ". . . whatever forces caused the water to seep into holes were beyond simple considerations of the depth of water-fillings, and by unpredictable amounts which could change in different ways depending on whether or not stabilized water drops were increased or decreased" (sic).

a. Which sentence immediately reveals an error or misconception?

b. Explain why the perk rate increases and decreases were not proportionate. Was this unpredictable?

c. If the "stabilized" rate had been measured not after 1 day but after 120 days, and the "average X" would have been very large, approximately what perk rate increase and decrease would have been noticed?

The answers are in Appendix D.

6.4.6 Plugging the Bottom of the Perk Hole

Weibel et al.[4] plugged the bottom of 4-inch-diameter holes with concrete, measured perk rates, and reported that the perk rate was not affected. They did not offer any explanations, but you could. (The explanations are in Appendix D.)

6.4.7 Effect of Perk Hole Diameter (or Radius)

Prove that:

a. When X is much smaller than h, i.e., $X < < h$, the level of water in a perk hole falls faster if the hole radius is smaller.

b. After true stabilization, when the "average X" is much larger than h, i.e., $X > > h$, the water level falls faster if the hole radius is smaller.

Proofs follow.

The volume of water inside a cylindrical test hole is given by $\pi r^2 h$. The rate of flow out of the hole is equal to the rate of decrease in water volume. And when $X < < h$, we have a situation similar to that in the can with perforated bottom and sides (see 6.4.4b); so, flow is proportional to a constant times the pressure times the wetted areas. Hence, the decrease in water volume in an instant of time is:

$$-d\,(\pi r^2 h)/dt = c\,p\pi r^2 + c\,p'2\pi rh$$

The pressure p on the bottom is h, and the pressure p' on the wetted sidewall of the hole is $\frac{1}{2}$ h. Therefore,

$$= ch\pi r^2 + ch\pi rh.$$

Rearranging terms and treating r as if it were a constant and h as a variable, and simplifying the equality, we get

$$-d\,(h/r)/dt = (ch/r)(1 + h/r)$$

From this expression, we can see that, if r is a constant and extremely large in comparison to h so that $h/r = 0$, the rate of decrease in h is given by $-dh/dt = ch$. This can be visualized as a hole 100 feet in diameter with a surface head of 6 inches: the initial infiltration rate will be proportional to the surface head. We can also see that, if r is extremely small in comparison to h, the term $(ch/r)(1+h/r)$ becomes extremely large, so that the rate of decrease in h is extremely fast.

When $X > > h$, or stated another way, when $h < < X$, the effect of p and p' on the gradient is negligible. We have a situation in which the average gradient is close to unity, and the velocity of flow is almost a constant, as seen in the previous chapter. Also, the rate of infiltration through the hole sidewalls and bottom is much lower than at the beginning of a test when X is zero; so, the hole sidewalls do not limit or control flow. The hole is like a "point" source. The soil is absorbing water at a nearly constant rate. Hence, the rate of decrease in h is equal to rate of flow divided by πr^2. Therefore, if r is reduced by a factor of 2, the instantaneous rate of decrease in h is quadrupled.

6.4.8 The Olivieri-Roche Correction Factors

These correction factors are derived from ratios of perk hole volume to perk hole wetted area. Let us derive a simple and easy formula to facilitate calculations. Let V = volume and S = wetted surface. Then,

$$V/S = (\pi r^2 h)/(\pi r^2 + 2\pi rh) = rh/(r + 2h)$$

Let us assume that the average h during a measuring interval is about 7.5 inches. The V/S of a 6-inch-diameter hole with h = 7.5 inches is:

$$rh/(r + 2h) = 3 \times 7.5/(3 + 15) = 1.25$$

The V/S of a 12-inch-diameter hole with the same surface head is $6 \times 7.5/(6 + 15) = 2.14$. The correction factor for converting mpi from a 6-inch-hole to a 12-inch-hole is 2.14/1.25. Then, if one measures a perk time of 30 mpi in a 6-inch-diameter hole, one would have measured approximately $30 \times 2.14/1.25 = 58$ mpi in a 12-inch-diameter hole, when the correction factor is applicable.

The Olivieri-Roche correction factors are not always applicable. Discuss the limitations to their applicability. (Hint: Review 6.4.7 and think about h and X. The answer is in Appendix D.)

REFERENCES

1. Conta, J.F., et al. 1985. Percolation tests, soil texture, and saturated hydraulic conductivity in lacustrine soils in North Dakota. *J. Environ. Qual.* 14: 191–194.
2. Hantzsche, N.N., et al. 1982. Soil textural analysis for on-site sewage disposal evaluation. In ASAE Pub. 1–82, pp. 51–60. ASAE, 2950 Niles Road, St. Joseph, Michigan 49085.
3. Winneberger, J.T. 1984. *Septic Tank Systems*. Butterworth Publishers, Stoneham, MA.
4. Weibel, S.R., et al. 1954. Studies on Household Sewage Disposal Systems. (Part III). Public Health Service Pub. No. 397.

5. "Committee." (1957, and revised in) 1967. Manual of Septic Tank Practice. Public Health Service Pub. No. 526.
6. Otis, R.J., et al. 1980. EPA Design Manual: Onsite Wastewater Treatment and Disposal Systems. EPA Report No. EPA-625/1–80–012.
7. Olivieri, A., and R. Roche. 1979. Minimum Guidelines for the Control of Individual Wastewater Treatment and Disposal Systems. California Regional Water Quality Control Board, 1111 Jackson Street, Room 6040, Oakland, CA 94607.
8. Van Kirk, J.L., et al. 1981. Percolation Testing for Septic Tank Leach Fields at Roadside Rests. Report FHWA/CA/TL/81/05. National Technical Information Service, Springfield, VA 22161.

7

Size of Leachline

As of now, whoever has had the fortitude to read through all the preceding chapters and solve the problems knows more about perk hole testing than at least 95% of perk testers. The rest of this book should be much easier.

Now that we know how to test perk holes, how do we translate perk rates into leachfield size requirements? I asked myself this question, and tried to answer it as if I were in front of a court of law. My answer had to make sense, and it had to agree with (and be endorsable by) other experts. Furthermore, it had to comply with the Uniform Plumbing Code,[1] which had been incorporated into (San Bernardino) County Code (by reference) in 1948.

The leachfield size requirements found in Table I-4 of the UPC (Uniform Plumbing Code) had no scientific basis:

1. They depend on the texture of the soil, and texture is not a reliable indicator of permeability. This lack of reliability had been known at least as far back as 1955.[2]
2. Even if texture were a reliable indicator, the textural classifications given in Table I-4 are so vague that they are of little use. (For instance, "sandy clay," "clay with considerable sand or gravel," "clay with small amount of sand or gravel"). It is bad enough to see, as I do, that almost everything is "silty sand" to experienced engineers, and that some designate a sand with only 2% clay as "clayey sand." To entrust the poorly defined UPC classifications to contractors, building inspectors, and sanitarians is bound to cause trouble.

3. From contractors, and from a sanitation district manager, I learned that leachlines (and replacements) installed in very sandy soils per UPC requirements invariably failed after 5 to 8 years. From some engineers, I learned that some leachlines had failed after only 2 years. I myself checked some leachlines that were installed per UPC standards in sandy clay and had failed after 4 months.

4. Colleagues in other California counties told me they had been confronted with so many failures that they had abandoned the UPC's Table I-4 requirements entirely.

5. As seen in the 1976 edition of the UPC, the authors of Table I-4 had the notion that clear water is absorbed at least 5 times faster than sewage, so the minimum size of leachfields was predicated upon this notion. The actual figure should be roughly 45 to 1,100 times faster. For instance, 1 ipm is equivalent to an absorption rate of 1440 inches per day and to 900 gallons of water/square foot/day. But after a clogging mat forms, the soil may absorb only 0.8 gallons of sewage/square foot/day. So, 900/0.8 = 1,100. The other figure, 45 times faster, derives from 0.01 ipm and a (clogged soil) absorption of 0.2 gallon/s.f./day.

   (The more recent UPC editions have eliminated the sentences implying that there is at least a fivefold difference between the clear water and the sewage absorption rates. Yet to this day I see perk reports prepared by otherwise reputable consultants, in which they make calculations based on an exact fivefold difference.)

6. I suspected that UPC's Table I-4 did not benefit from all the expertise available at the time it was prepared, and my suspicion turned out to be correct.*

---

*I asked a most reliable source, "How could an expert committee have cooked up UPC's Table I-4, obviously without the benefit of research or even well-known published soils data?" His answer was as follows. Not all of the members of the committee were experts, so they invited a well-known expert (I cannot reveal his name) to help them. This expert requested a fee for his services. The members of the committee, who are and must be unpaid volunteers, could not grant a fee. Hence, the expert did not participate. But the committee had to come up with recommendations, and so it did.

Obviously, the UPC could not be a source of valid leachfield size requirements. However, the EPA had recently conducted further studies to define the perk methodology; and even better, these studies had been evaluated and criticized by an "outsider" expert—Winneberger. (In the scientific community, a study or publication is not considered to be reliable until it has had the benefit of exposure and criticism—the more, the better.) So, here was information I could use.

7.1 PRACTICAL SIZE OF LEACHLINE

In the late 1970s, the National Environmental Health Association complained to the EPA that the old USPHS Manual of Septic Tank Practice (USPHS Pub. No. 526) was far from adequate. The EPA commissioned (indirectly) Otis et al. to check and standardize traditional perk procedures and soil absorption area requirements, and to incorporate their findings into a manual on onsite disposal of sewage. The result was the EPA Design Manual.[3] I took data from this Manual's Table 7.2, converted sewage application rates from gallons/s.f./day to square feet (of absorption area)/gallon/day (because the UPC defines absorption area in this way), and plotted the results on semi-log paper. I did the same thing with the absorption area requirements recommended by Winneberger[4]; but I plotted half the square footage required so the data would fit on the same sheet of semi-log paper. This is shown in Figure 7.1. In Figure 7.2 I compared the EPA's and Winneberger's recommendations for the 0- to 60-mpi range. Since Winneberger's recommendations might have been influenced by his use of 4-inch-diameter perk test holes, I used the Olivieri-Roche hypothesis to calculate and plot the effect of differences in diameter. The effect was not pronounced.

Surprisingly, EPA and Winneberger are off only by a factor of about 2 (in the 1- to 60-mpi range). This is close, and in excellent agreement, considering all the variables involved in

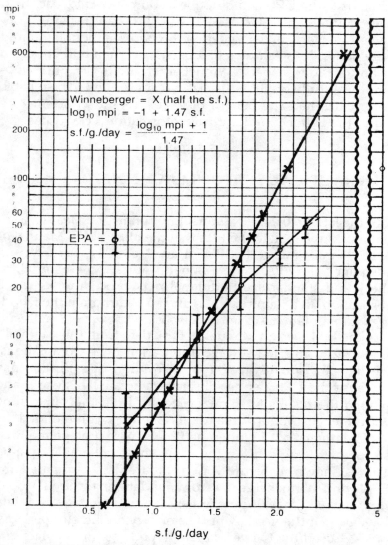

Winneberger = X (half the s.f.)

$$\log_{10} mpi = -1 + 1.47 \text{ s.f.}$$

$$\text{s.f./g./day} = \frac{\log_{10} mpi + 1}{1.47}$$

EPA = ⊘

s.f./g./day

Figure 7.1 EPA's and Winneberger's recommendations for absorption area (square feet per gallon of sewage per day) versus measured perk times, mpi. The x's represent the Winneberger absorption areas divided by 2, and the circles represent the midpoints of the EPA absorption area ranges; the ranges are indicated by vertical lines.

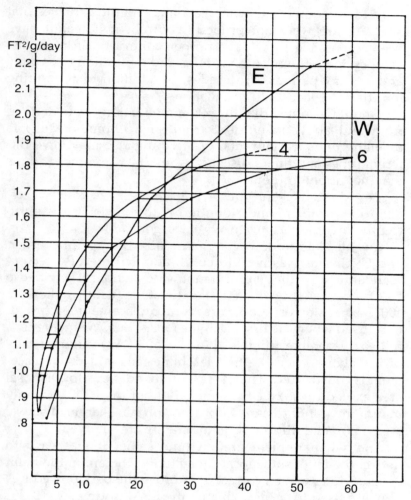

Figure 7.2 Effect of hole diameter on Winneberger's data. The E curve goes through the midpoints of the EPA ranges shown in Figure 7.1. The W 4 curve is a plot of the Winneberger data in Figure 7.1; it might reflect results obtained in a 4″-diameter perk hole. The W 6 curve is the W 4 curve modified to show results obtained in a 6″-diameter perk hole.

EPA's derivation of experimental data. (All the variables that affect failure were not properly controlled. It would have been impossible to achieve proper control of effluent discharges and effluent compositions and infiltration rates daily and at peak periods for decades, in addition to accounting for effects of leachfield installations.)

It should be noted that the recommendations were derived from and apply to effluent from single homes. Multiple dwelling units, restaurants, or anything else may have different types of effluent discharges and the recommendations may not apply to them.

The expected useful life of a leachline is 20 to 50 years, according to Otis;[3] and "long term" to forever if it is sufficiently large, according to Winneberger.[4]

Now all I had to do was to harmonize EPA, Winneberger, and UPC. This was not difficult: I adopted the EPA recommendations, as they are a nationwide standard. I added a UPC requirement which I had to abide with anyhow—a 100% expansion or reserve area for a replacement leachfield, to be used when the first leachline fails. Thus, Winneberger's full-size leachline (double the size of EPA's) automatically comes about when the first leachline fails and the replacement is constructed. The EPA recommendations for disposal area exceed those of UPC's Table I-4, but the UPC sets only minimum standards, and does not prohibit larger disposal areas. So, all authorities' requirements are satisfied.

(And when implementation time came, it was easy to deflect the ire of some local consultants who are in the habit of squawking when I correct or improve the perk procedures they are used to. I said, "Blame the EPA, not me.")

However, I made one modification to the EPA recommendations, with the approval of local perk consultants. This came about because one of them pointed out a problem. He had tested two small lots. One lot was suitable for septic systems, the other one was not. The lots were identical, except that the average perk time of one was 5 mpi, and of the other, 6-mpi. Only 1 mpi difference caused one to be rejected because the EPA requirement jumped from 0.83 to

1.25 square feet per gallon per day: a 50% increase in leach-field size. The 6-mpi lot did not have enough area to accommodate this increase. The perk tester felt he could not explain this to the owner of the rejected lot. The "jump" was odd, and was perceived as unfair. So, the solution was to abandon the "jumpy" EPA requirement ranges and to use the smooth curve which goes through the midpoint of these ranges shown in Figures 7.1 and 7.2.

This is terribly unscientific, as this curve implies that a high precision exists. But for the sake of fairness, one must make allowances.

REFERENCES

1. IAPMO. Uniform Plumbing Code (1976 to 1985 eds.). Los Angeles, CA.
2. SERL. 1955. Studies in Water Reclamation. Sanitary Engineering Research Lab. Tech. Bull. No. 13, I. E. R. Series 37, University of California, Berkeley.
3. EPA (Otis, R.J., et al.) 1980. Design Manual: Onsite Wastewater Treatment and Disposal Systems. EPA-625/1-80-012.
4. Winneberger, J.T. 1984. *Septic Tank Systems*. Butterworth Publishers, Stoneham, MA.

8

Factors Affecting Failure
of Leachlines

Leachlines can fail in two ways: They can fail to absorb the septic tank effluent, and/or they can fail to treat the effluent (i.e., filter out parasites and pathogenic microorganisms). This chapter addresses the first type of failure.

Failure to absorb the septic tank effluent can be ascribed to various variables, some of which are interrelated:

Amount and frequency of effluent discharge
Composition of effluent
Type of soil, vegetation, and climate
Amount and configuration of absorption surface
Installation
User habits and behavior
Time (age)

8.1 AMOUNT AND FREQUENCY OF EFFLUENT DISCHARGE

Regarding household sewages, perhaps the most important variable influencing failure is the amount of effluent discharged per square foot of leachline absorption area. In addition to possibly exceeding normal soil infiltration rates, excessive loading of a leachline promotes anaerobic conditions and formation of a less-permeable biomat. Experienced septic system installers (M. Fernandez, D. Arnegard) told me that about 10% of the failures they encounter, occur in "young" leachlines affected by leaky plumbing (usually the toilet). After they fix the leak and install flow reduction devices, the leachline becomes functional again. Siegrist et al. suggest that flow reduction can translate to reduction in absorption area.[1]

According to Hargett et al.[2], the frequency of discharge (or dosing) presents little advantage if the discharge rate (loading rate) is high; if the loading rate is low, dosing so as to cause no ponding on the leachline bottom helps maintain more aerobic conditions and hence higher infiltration rates (more permeable biomat).

Asbury and Hendrickson's[3] experiences in New Mexico are not new to southern California counties. They found out that it is just too much bother to promote aerobic conditions by means of aerobic pretreatment of septic tank effluent: too much neglect by users and too many malfunctions. The traditional septic system is less costly and less trouble-prone.

Otis[4] noted that if septic tank effluent is treated aerobically before it is discharged to the leachline, clogging may be retarded but not avoided.

All in all, it is important not to overload a leachline, and not to exceed the loading rates suggested in the previous chapter. And in order not to exceed such rates, one should keep in mind the type of information presented in Table 8.1. That is, average sewage flows ("interior" usage) may vary by a factor of (almost) 3 between different California regions, and by who knows how much between different localities or neighborhoods. Therefore, a septic system designer should never follow the common practice of using discharge figures found in a reference book (e.g., Metcalf and Eddy). These figures tend to be gross averages. This means that, even if the averages are accurate, half of the septic systems based on these averages are overdesigned, and half are underdesigned (and this is a problem). Furthermore, such averages are not specific to particular neighborhoods, localities, or ways of life. A septic system designer should obtain information about residential sewage discharges from local water

Table 8.1 Water Consumption in California (Adapted from Ingham)[5]

Hydrologic Study Area	Gallons/Capita/Day		
	Residential	Interior	Exterior
North Coastal	354	198	156
San Francisco Bay	122	68	54
Central Coastal	132	73	58
South Coastal	122	68	54
Sacramento Basin	239	133	105
Delta-Central Sierra	214	120	94
San Joaquin Basin	296	165	130
Tulare Basin	247	138	109
North Lahontan	334	187	147
South Lahontan	207	116	91
Colorado Desert	257	144	113
Mean	229	128	101

and/or sewer districts; and about specific commercial/industrial sewage discharges, from actual meter readings at existing similar projects, if possible. Table 8.1 refers to water use in the late sixties. Since that time, California and the rest of the United States came to know the Jacuzzi, the monster-sized tub, and flow-restriction devices. Immigrants of various nationalities and also the natives' new ways of life have added variety and spice to sewage flows. Therefore, one should be wary of published flow data.

8.2 COMPOSITION OF EFFLUENT

Table 8.2 serves to illustrate that what is true of sewage flows (i.e., variability) is also true of sewage flow constituents. Even the different studies' means vary considerably.

BOD means biological oxygen demand. If we measure the oxygen consumed by microbes when a bit of sewage is placed in a relatively large water volume and incubated 5 days, we measure BOD_5. The more biodegradable matter there is in the sewage, the higher its BOD.

COD is chemical oxygen demand. Whatever can be oxi-

Table 8.2 Summary of Effluent Data from Various Septic Tank Studies (Source: EPA[6])

Parameter	Source				
	7 Sites	10 Tanks	19 Sites	4 Sites	1 Tank
BOD$_5$					
Mean, mg/l	138	138[a]	140	240[b]	120
Range, mg/l	7–480	64–256	–	70.385	30–280
No. of Samples	150	44	51	21	50
COD					
Mean, mg/l	327	–	–	–	200
Range, mg/l	25–780	–	–	–	71–360
No. of Samples	152	–	–	–	50
Suspended Solids					
Mean, mg/l	49	155[a]	101	95[b]	39
Range, mg/l	10–695	43–485	–	48–340	8–270
No. of Samples	148	55	51	18	47
Total Nitrogen					
Mean, mg/l	45	–	36	–	–
Range, mg/l	9–125	–	–	–	–
No. of Samples	99	–	51	–	–

[a]Calculated from the average values from 10 tanks, with 6 series of tests.
[b]Calculated on the basis of a log-normal distribution of data.

dized by chemical means rather than by microbes is considered to be COD. COD is equivalent to "BOD plus nonbiodegradable oxidizable matter."

Suspended solids (SS) comprise anything suspended in the sewage: bacteria, fecal particles, emulsified oil, etc.

Both BOD and SS promote clogging mat formation. BOD provides "food" to microorganisms which grow and decompose or secrete clogging substances. Some SS particles may plug the soil pores directly; some may be part of BOD and stimulate microbial activity.

At one time it was believed that adjustments to absorption area could be made on the basis of an empirical formula, the cubic root of (BOD+SS)/250. But this formula did not survive scrutiny.[7]

It should be noted that BOD and SS are convenient words

used to relate "something" to clogging. They convey no information as to specific composition and as to how "clogging" they are. For instance, emulsified oil and grease might not be a large part of either BOD or SS, yet they are highly clogging.

The effect of oil and grease can be pronounced. The washings from restaurant kitchens contain quite a bit of oil and grease. At least in theory, the oil and grease are captured by means of a grease interceptor (or trap), so that they do not plug up sewer pipes and do not overload the septic tank. According to J. and M. Fernandez (local septic system contractors), they solved a failure problem by deviating from code requirements, as follows. A restaurant had a 5,000-gallon septic tank discharging into one seepage pit, and a 1,000-gallon grease interceptor through which the kitchen effluents discharged into another seepage pit (of the same size as the septic tank pit). The interceptor's pit failed yearly and had been replaced every 2 years, for 8 or 10 years. They checked the interceptor's influent and effluent, and found out that the temperature and composition of these were about the same; i.e., the grease was not being trapped. So, they directed the flows that used to go to the septic tank to the grease interceptor, and vice versa. The septic tank's larger volume increased retention of grease. Eight years after the switch, everything is still working fine.

Waters with a high sodium/(calcium + magnesium) ratio tend to deflocculate (disaggregate) expanding clays, which can seal soil pores. But at least in the common types of soils used for leachfields, high sodium, and normal use of soap and detergents and bleach have not resulted in noticeable impairment of leachfields.[6]

8.3 TYPE OF SOIL, VEGETATION, AND CLIMATE

As discussed in previous chapters, the type of soil influences how much water moves, how fast, and to where. Vegetation and climate influence evapotranspiration.

The interrelationships between soil, vegetation, and climate are rather special and may present problems not yet described in the septic systems literature.

For instance, in the semiarid San Bernardino Valley, in a community where single houses are on 7,200-square foot lots, leachline failures were frequent after heavy rains. Roof runoff had nowhere to go but to the leachfield area. (Seepage pits did not fail.)

Most soils in the densely forested mountain resort community of Fallsvale (San Bernardino County, California) are very coarse textured. Most of the rainfall falls in winter; the soils are fairly dry in summer and fall. Conifers and oaks are common. The conifers, and particularly the "incense cedar" (*Calocedrus recurrens*), grow roots toward and into leachlines. (And what roots they are! Try to imagine a plug of densely packed fibers within a leachline pipe.)

Mr. W. Fagerstedt has been installing and repairing septic systems in Fallsvale for over 40 years, and has acquired unique insights into local septic system problems. He has kindly agreed to share some with us. He noticed that roots plug leachpipes severely after only 2 to at most 10 years of leachline use. (The first symptom of this condition is that toilets cannot be flushed, tubs cannot be drained.) In order to minimize this problem, he suggested shorter and wider leachlines. His prescription was a leachline 40 feet long and 4 feet wide, with 4 feet of gravel below the pipe. Although the absorption area provided by this leachline is generally smaller than that required per criteria discussed in the previous chapter, it does not make any sense to lengthen a leachline even to 1,000 feet if it gets plugged up near its inlet.

By the way, downstream communities obtain high-quality water from the Fallsvale area. Therefore, large-scale use of copper sulfate or rock salt (root killers) could degrade the water and is not a viable solution. The clearing of trees in an area famous for its scenic forest is not a viable solution either.

Mr. Fagerstedt also noticed that the problem is much worse in leachlines that are used intermittently by weekend users or vacationers. The roots "search" for moisture where it is available, and grow into the pipe. In this regard, he also noticed that gravel drains easily and helps keep the roots down at the bottom of the trench; however, slag has a rough surface which retains moisture droplets and attracts roots. On occasion his backhoe had trouble digging failed slag-filled leachline trenches which had become a solid mass of roots and slag.

Another way Fagerstedt mitigates the root problem is by constructing shallow (6-foot-deep) seepage pits with unperforated blocks and no gravel. The roots grow outside, around and just below such pits; and, from what I have seen, the roots must be absorbing a very large amount of water from the sides and bottoms of those pits (one has been in continuous use for 20 years, according to Fagerstedt). And yet another way is by installing unperforated rather than perforated pipe in the portions of a leachline that are near trees.

A Wrightwood (San Bernardino County, CA) septic system installer (J. Freitas) told me he traced a root from a plugged leachfield to a cottonwood tree about 40 feet away. Elsewhere, in a more arid area, I traced a 4-inch-diameter root for almost 40 feet to a scrub oak bush.

8.4 AMOUNT AND CONFIGURATION OF ABSORPTION SURFACE

In the previous chapter we arrived at figures for the amount of absorption surface per gallon of effluent. But now, what is the most efficient shape for this absorption surface? Should it be a narrow deep trench or a broad shallow trench?

The EPA Manual's[6] absorption area requirements are mostly derived from experiences in the north central states. To find out what kind of leachline configuration is used there, I read Reference 8 and called its author. Machmeier kindly provided the following information (personal communication, 7/3/84) regarding practices in Minnesota: They use 2- to 3-foot-wide leachline trenches, maximum 30 inches deep, with 12 inches of gravel below the pipe. (Under special conditions they allow up to 2 feet of gravel below the leach-pipe, and a cover of up to 3 feet of soil; but this is "not recommended.") Leachlines (i.e., microbes) need oxygen, and therefore are shallow. Also, they try to pond the effluent and maximize biomat formation, particularly so in soils with perk times of less than 5 mpi (to promote unsaturated flow and purification of effluent). This is accomplished by direct-ing all the flow to a relatively short leachline; when this leachline is flooded, the excess flow goes out through a drop box and goes into another leachline, and so on sequentially. Some leachlines have been in use for more than 45 years and are still working.

Bowman has reported virtually no failures in septic sys-tems which have been in use for about 30 years.[9] These sys-tems serve three-bedroom houses, and each consists of 400 lineal feet of shallow leachline, 2 feet wide, with 1 foot of gravel below the pipe. This is equivalent to about 1600 square feet of absorption area, about half of it sidewall, and half bottom. Since the soils have perk times of 5 to 25 mpi,[9] and assuming an effluent discharge of 300 gallons/house/day, only about 900 square feet of absorption area would be required according to EPA.[6] Hence, the longevity reported by Bowman is not surprising.

On the other hand, studies in Oregon support the view that leachlines deeper than 36 inches are advantageous because of "increased absorptive surface, increased hydrau-lic head [actually, surface head h] . . ." and other factors.[10]

Winneberger and Klock[11] are partial to sidewall absorption area. After the bottom of the leachline is clogged by migrat-ing clay or biomat, the sidewalls keep on absorbing the sew-age.

As far as I can see, narrower (than 2 feet), deeper (with more than 2 feet of gravel below the perforated pipe) leach-line trenches absorb more efficiently than broad shallow ones, on a per-square-foot basis. Also, they are less likely to be affected by high rainfall (or by watering the lawn). However, narrow trenches might be more likely to be affected by roots, especially in areas where moisture is scarce during part of the year.

Regarding longevity of leachlines, configuration is less important than the amount of absorption surface. The less-efficient configuration of a wider, shallower leachline (with higher ratio of bottom to sidewall area) can be compensated by increasing leachline length. Having enough absorption surface is the paramount consideration.

It should be noted that in the common range of soils used for septic systems (1 to 60 mpi soils), the main factor limiting absorption is the permeability of the biomat. The amount of absorption area required is fairly independent of the type of soil. (A 60-mpi soil requires only three times more absorption area than a 1-mpi soil.) This relative independence has been noted in the literature.[12] Also, a local knowledgeable septic system contractor looked up his voluminous repair records and concluded that regardless of the kind of soil the leachlines were in, if the absorption area was less than 550 to 600 square feet, the frequency of failures increased drastically (M. Fernandez, personal communication).

8.5 INSTALLATION

A septic system might be well-designed, but it may fail fast if the installer is not knowledgeable and conscientious. I have noticed that many contractors just "go through the motions." They dig the leachline trenches, put gravel and pipe and paper cover, pass the inspection, and fill the top of the gravelled trenches with excavated soil. Often they dig when the soil is moist, and end up compacting the soil so badly that its absorption capacity is drastically reduced. Sometimes the gravel has much silt and clay, either because it comes that way or because it gets mixed up with local soil when it is pushed into the trench. The silts and clays, or "fines," are washed downward and plug the soil pores at the bottom of the leachline. Sometimes, after the trenches are dug, rainfall washes out fines from the trench sides and forms a fairly impermeable crust at the bottom of the prospective leachline; then, the contractor dumps gravel on top of this crust, and proceeds to completion.

If the top of the leachline is not compacted with excess soil backfill, the soil settles in about one year and leaves a depression over the length of the leachline. Rain can accumulate in this depression, and can flood the leachline trench below. If too much backfill is left on top and the leachline is on sloping ground, the mound of backfill may trap natural drainage and direct it into the leachline, or it may divert it into paths where it can cause erosion over other leachlines.

8.6 USER HABITS AND BEHAVIOR

After long holidays, a not uncommon sight in our local resort communities is a vacation cabin or house with a failed (and smelly) leachfield, and, next to the street, a pile of trash bags overflowing with paper plates, empty beer cans, and other signs of the good life. A local septic system installer has corroborated that there is a relationship between failure and the accumulated trash: He gets many calls after homeowners have entertained a dozen or so of their friends and family during the holidays (J. Freitas, personal communication).

At the higher end of the economic scale, wealthy residences tend to have large bathtubs and large per capita discharges. Yet at the opposite end of the scale, when I inspected a failed leachfield in front of a little house, I noticed that about a dozen people lived in it. If each flushed the toilet only twice a day . . .

Proper design must take into account the type of neighborhood and lifestyle. Never should a responsible septic system designer size a leachfield solely "by the book."

Other user behaviors that affect septic system performance are: no pumping of septic tank, discharging motor oil down the drain, and postponing repair of leaky plumbing fixtures.

8.7 TIME (AGE)

Do leachlines always have a finite life? There are no data to support either a "yes" or a "no" answer. The question itself is not a very good one, as we have not specified parameters such as amount, frequency, and composition of sewage flows, and amount of absorption area, type of soil, and climate.

My own view is that, over most of the United States, they do. But this finite life could be extended to hundreds or thousands of years, if the leachlines were built a few times larger than usual.

We can assume that the fate of the biomat is similar to that of (natural) soil organic matter, if the leachline is abandoned for a long period. In tropical climates, organic matter usually does not accumulate in the soil. It is almost totally decomposed. In cold climates, and particularly under wet (anaerobic) conditions, organic matter accumulates (and forms organic soils).

Under hypothetical "average"* local conditions, if the SS discharged into a leachline are not decomposed at all, it would take about 100 years for the SS to completely fill the spaces between the gravel and cause failure (due to lack of storage volume within the leachline). Most of the local septic system installers I contacted have not noticed any such accumulation when they have dug old leachlines, but one (M. Fernandez) told me that he saw about 2 inches of a whitish material over the bottom of an old abandoned leachline. I

---

*Average = A house discharges 300 gallons/day with 200 mg/liter of SS into a leachline with 9,000 liters of gravel void space (or empty space between the gravel particles). The densities of SS and of water are assumed to be equal; otherwise, the SS would not be in suspension.

would not worry too much about SS accumulation within leachlines, if the septic tank is pumped as needed.

8.8 MISCELLANEOUS

E. Scheider has been installing (and monitoring the performance of) septic systems for 30 years in and near the community of Wrightwood. Some of his observations are of interest to us.

As he tells it, he has found six failed leachlines in which black slime (mainly from SS and BOD discharges) filled all the voids of the gravel below the leachpipe; in about half of these cases, this condition was evident only near the beginning of the leachpipe. In one case the whole length of the leachline was affected. In each one of these cases the two compartments of the septic tank were full of sludge. The worst case, where the whole leachline length was affected, had a failure 1 year after a water softener was installed; he found all the baffles of the steel septic tank had been corroded and had practically vanished. He thinks the water softener salts killed the bacteria which digest organic matter within the tank. (This explanation is not supported in the literature. Perhaps the sodium from the salts or from the lye common in drain openers deflocculated the organic colloids in the tank, or caused them to "bulk"). Also, he believes failure problems are worse in "weekender" cabins, because the bacteria starve out during long periods of nonuse.

8.9 A WORD TO THE WISE

With 8 years of coursework in soil science behind me, and after having read a ton of publications concerning septic systems, I still don't have all the answers about leachline longevity.* No one has.

---

*False modesty is one of my virtues.

I have learned much from contractors who are not satisfied with "just meeting the code," and who experiment, observe, and keep track of what they do and of the consequences. In practical matters, I suspect they are more knowledgeable than the average perk professional. Perk professionals would do well to consult with such contractors before designing septic systems.

REFERENCES

1. Siegrist, R., et al. 1978. Water conservation and waste-water disposal. ASAE Pub 5-77:121–136.
2. Hargett, D. L., et al. 1982. Soil infiltration capacities as affected by septic tank effluent application strategies. In: Proceedings of the 3rd National Symposium on Individual and Small Community Sewage Treatment. ASAE Pub. 1–82, pp. 72–84. ASAE, P.O. Box 410, St. Joseph, Michigan 49085. (See ref. 4 for recent address.)
3. Asbury, R., and Hendrickson, C. 1982. Aerobic on-site systems studied in New Mexico. *J. Environ. Health* 45: 86–87.
4. Otis, R. Soil clogging: Mechanisms and control. 1985. In: Proceedings of the 4th National Symposium on Individual and Small Community Sewage Treatment. ASAE Pub. 07–85, pp. 238–250. ASAE, 2950 Niles Road, St. Joseph, Michigan 49085–9659.
5. Ingham, Alan. 1980. Residential Greywater Management in California. (California) State Water Resources Control Board, P.O. Box 100, Sacramento, CA 95801.
6. EPA (Otis, R., et al.). 1980. Design Manual: Onsite Wastewater Treatment and Disposal Systems. Report No. EPA-625/1-80-012.
7. Otis, R., et al. 1977. On-site disposal of small wastewater flows. Department of Civil and Environmental Engineering, University of Wisconsin, Madison.

8. Machmeier, R. E. 1981 Revision. Town and Country Sewage Treatment. University of Minnesota, Agricultural Extension Service Bulletin 304. St. Paul, Minnesota 55108.

9. Bowman, J. O. 1982. Reliability of on-site sewage disposal systems in Fairfax County, Virginia. *J. Environ. Health* 44:249–252.

10. Ronayne, M. P., et al. 1982. Final Report, Oregon On-Site Experimental Systems Program. Oregon Department of Environmental Quality, P.O. Box 1760, Portland, Oregon 97207.

11. Winneberger, J. T., and Klock, J. W. 1973. Current and Recommended Practices for Subsurface Waste Water Disposal Systems in Arizona. ERC-R-73014. Engineering Research Center, Arizona State University, Tempe, Arizona 85281.

9

Size of Seepage Pits

I doubt that there is a reliable study relating seepage pit absorption area to precise years of use before failure. The many variables involved would make any results either inconclusive or else valid only for a particular locality or restricted set of conditions.

The performance of seepage pits is affected by almost the same variables affecting leachlines (see Chapter 8), plus a few others: small to large surface head, stratification of soil profile, position of porous strata within this profile, lateral extent and continuity of strata, and angle of such strata with the horizontal.

Three relevant quotations from the USPHS Manual follow:[1]

1. "It is important that the capacity of a seepage pit be computed on the basis of percolation tests made in each vertical stratum penetrated." [The test at each stratum is performed just as for leachlines.]

2. "The weighted average of the results should be computed to obtain a design figure. Soil strata in which the percolation [times] are in excess of 30 minutes per inch should not be included in computing the absorption area."

3. "Although few data have been collected comparing percolation test results with deep pit performance, nevertheless the results of such percolation tests, while of limited value, combined with competent engineering judgment based on experience, are the best means of arriving at design data for seepage pits." [In other words, the Manual's authors are saying, "We don't know."]

The EPA Manual[2] prescribes the same methodology, but it omits the cautionary statements and the admission of ignorance. This is most unfortunate.

Step by step, let's see what the problem is.

Let us visualize a pit in a uniform soil. With time, the clogging mat will progress upwards; the column of sewage within the pit will rise and be absorbed in the as-yet-unclogged upper portions of the pit. But as this sewage column increases in height, so does the surface head, and hence absorption through the clogged portions of the pit also increases.

Now, if the soil is stratified (like, for instance, horizontal layers of permeable sand sandwiched between layers of impermeable clay or caliche), a percolation test on a sand stratum may grossly overestimate the amount of absorption of sewage. The saturation of the soil during the test period is short-range, but after a pit has been in use for some time, the effluent moves horizontally over the clay stratum until it can infiltrate (very slowly) over a large area of this stratum. That is, the path of flow within the soil (the X) increases, and the resulting gradient is much smaller than under test conditions. Without a substantial surface head, little effluent movement and little absorption might take place.

Also, if two pits have the same amount of permeable and impermeable strata, but one has the permeable strata concentrated toward the bottom, this pit will derive more benefit from the same amount of surface head.

If the strata are tilted toward the vertical, lateral movement from the upper sidewalls might sometimes be restricted, but the movement downward will be enhanced.

When pits are installed in series, all the flow goes to the first pit, and the overflow goes to the second, and so on. If a pit is the first one in a series, and the soil is stratified, we do not have a seepage pit but an injection well. In this case, the weighted average of perk rates has no meaning at all. Sewage will not be absorbed in any manner even remotely related to the rates given in Chapter 7 for leachlines. We just don't know what kind of gradients or paths or biomat permeability rate changes we are dealing with.

There is some merit to the second USPHS Manual statement, that (strata with) perk times in excess of 30 minutes per inch should not be included in computing the absorption area. In the range of 1 to about 20 minutes per inch, soils drain easily through macropores. The number of macropores and the ease of drainage decrease when the mpi increases: the smaller pores or capillaries retain the water more tightly. Soils with perk times higher than 60 mpi tend to be high in clay and are prone to such problems as erratic measurements, easy compaction or smearing of sidewalls, poor drainage, and relatively poor absorption.

As for the third of the USPHS statements, it is easy to agree with its first part—that "such percolation tests" are of "limited value." But as for the use of "competent engineering judgment," I am amazed at its optimism. I have seen very few instances in which consultants did exceed minimum legal requirements, even when unusual soil conditions made it imperative to exceed them. Sanitarians from other jurisdictions have told me of similar experiences.

Winneberger lamented the fact that "minimum requirements become the standards of practice."[3] It is not difficult to understand why. Some perk report preparers are afraid to antagonize their clients if they propose anything that is not "minimum" and costs money or lowers profit. For instance, not long ago, I reviewed a perk report, and cut by half the density therein proposed for a residential project. The pits were to be located in marginal soils. The calculations in the report agreed with minimum requirements, but taking into account the soil strata, the local history of perched water tables, and so on, a high density was unjudicious. The engineer who prepared the "All's OK" perk report told me in private that he agreed with my action and felt relieved by it. Another perk report preparer confided that after (some of) his perk reports were not approved (by myself, as a county sanitarian), his clients didn't pay him his fees.

9.1 SIZING PITS THROUGH STANDARD LEACHLINE TESTS

Let us assume that we have a uniform soil all the way down to 50 feet. We test it as for leachlines, and from Figure 7.2 we determine that the absorption area should be, say, 1 square foot per gallon of effluent per day. If the pit is to serve a house discharging 300 gallons of effluent per day, the pit's absorption area must be 300 square feet. If the pit diameter is to be 5 feet, the perimeter is 5π, or 15.7 feet. Then, 300/15.7 = 19 feet deep. (The bottom area is never counted.) The 5-foot-diameter pit should be 19 feet deep *below inlet*. The inlet is usually 2 to 4 feet below grade. Therefore, the pit should be excavated down to about 21 or 23 feet below ground level.

Now let us assume that the soil is stratified. The thicknesses of the horizontal strata (below inlet) and their respective mpi (measured per standard leachline procedure) are as follows:

feet	mpi	feet × mpi
(6)	(40)	(not applicable)
10	20	200
5	10	50
(5)	(60)	(not applicable)
10	1	10
25		260

(We ignore completely any strata with mpi higher than 30.) Now, the weighted average mpi = 260/25 = 10 mpi for a pit that has an *effective* depth of 25 feet below inlet and a *total* depth of 25 + 6 + 5 = 36 feet below inlet. Now, from Figure 7.2, 10 mpi corresponds to 1.25 square feet/gallon/day. Since the effective sidewall area of the 5-foot-diameter pit is 5π × 25 = 390 square feet, we have that 390/1.25 or 313 gallons per day is the pit's supposed absorption capacity, for design purposes. In passing, this procedure might be fine in states where the maximum depth of pits is 10 feet below ground surface. Locally, the pleasures of testing little holes at 30 or 40 feet below ground have been greatly diminished by OSHA regulations.

9.2 THE FALLING HEAD TEST

In 1976, a committee of local perk consultants adopted a formula and a procedure to test and size seepage pits. Most unfortunately, this inaccurate methodology has spread to other California counties. To help contain the damage, I'll grant this methodology a bit more space than it deserves.

Briefly, the falling head test methodology is as follows: drill a 6- to 8-inch hole down to prospective seepage pit depth, fill it with water, and proceed more or less as if it were a leachline test but with the water level 10 to 40 feet above the bottom of the hole. The details are given in Appendix I.

The total volume of water absorbed in a measurement interval is easily calculated from the hole diameter and the fall in water level during the interval. The total area through which absorption takes place (inside the hole) is calculated from the hole diameter and from the average height of the water column (during the measurement interval). Now, dividing the gallonage absorbed by the average absorption area in square feet, we get "absorbed gallons/square foot/ time interval." Finally, multiplying by a factor which extrapolates the absorption during the measurement interval to absorption during a 24-hour day, and which takes into account that the clear water used in the test *allegedly* is absorbed 5 times faster than sewage effluent, we obtain the falling head formula,

$$Q = FD9/Lt$$

where

Q = gallons of sewage per square foot of pit side-wall

F = fall of water column during interval, in feet
D = diameter of hole, in feet
L = average height in feet of water column during measuring interval
t = time interval, in hours

The best feature of this formula is its worst: it is simple to the point of oversimplicity. But it looks respectable, so it has been widely used as a substitute for "competent engineering judgment." I myself might have played a role in making it look respectable, after I developed ancillary format and formulas. I do feel guilty. At least I tried warning perk consultants with a note, "The (falling head) formula means nothing. Results require interpretation." The warning had no effect whatsoever.

What is wrong with this formula is that:

1. It assumes that water is absorbed exactly five times faster than sewage. The justification for this absurdity is the 1976 edition of the Uniform Plumbing Code. As we saw in previous chapters, it is way off the mark, and the UPC never stated it was five times faster, anyhow.

2. Since the UPC allows use of soils with presumed absorption rates of 0.83 gallons of sewage per square foot per day, many pits have been installed in low-permeability soils which have Q's as low as 1.1 to 0.83, with predictable consequences. (Anyone who had made a simple calculation would have discovered that 1.1 gallons/square foot/day translates to 164 mpi!) An acquaintance's pit (about 5 feet in diameter and 40 feet deep) in a stratified soil rated for 1.1 gallons/square foot/day lasted 4 months; the replacement pit was dug even deeper and lasted 5 months.

3. The drilling often compacts the hole sidewall so badly that the soil permeability is decreased drastically. If this decrease were to occur consistently and uniformly, it could be corrected with an empirically derived fudge factor. But

this cannot be done. The type of drill, the eccentricity of the drill, the depth of soil, the type, density, and moisture of soil horizons, and other factors are responsible for variable results.

4. Its results require much interpretation. For instance, about three years ago a large-scale housing developer and his septic system contractor, R. K., asked me to allow them to deviate from the requirements specified in a perk test report. I examined this report. It was done "by the book," and the consultant who prepared it was as able as any.

The results per the falling head formula indicated that a three-bedroom home should have a 25-foot-deep pit (if I remember correctly). But R. K. had dug a few pits, and found out that the strata of impermeable clay and very permeable sand were very randomly distributed. In some cases, the 25-foot depth of the pit would have cut mostly through sand strata, and in other cases, mostly through clay strata. In the first case, the absorption area would have been more than adequate for a 10-year lifespan, and in the second, less than adequate.

So R. K. requested permission to deviate from the 25-foot depth requirement, and to dig the pits the way he had always done it (with satisfactory results). And this was as follows. No matter what the perk reports or engineering specifications were, he would dig until he had gone through *at least* 10 feet of sand strata. So, some of his pits were to be less than 25 feet deep, and some maybe 40 feet, or deeper. (The groundwater table, at 200 feet, was not a problem.) Here I was talking to someone who knew what he was doing. After I inquired with county building inspectors about his reputation and "workmanship," I agreed wholeheartedly to his proposal.

(I asked the developer how much he had paid for the perk test report. His answer was $27,000.)

5. If the falling head formula is to have any meaning at all, it should be $Q = 45FD/Lt$ (gallons of *water*/square foot/ *extrapolated* 24-hour day), or "pit mpi" = $20Lt/FD$. The "pit mpi" serve to give an approximate idea of the type of soil. In coarse-textured, permeable soils, the "pit mpi" are roughly similar in value to mpi values from leachline perk tests.

9.3 LONGEVITY AND ABSORPTION CAPACITY OF SEEPAGE PITS

In sizing seepage pits, there is really no formula or methodology that can substitute for good judgment. Good judgment must be based on theories and on perceptions about how such theories are (or are not) pertinent to a given set of soil conditions. My theories and views are presented below.

When a clogging mat forms in a leachline, absorption through the, say, 1-inch-thick layer of clogged soil may be reduced to about 0.2 gallons/square foot/day. If we increase the surface head suddenly by a factor of 10, the absorption rate might also increase by up to a factor of 10 for some time (see Chapters 5 and 6). Thereafter, at constant head, and more and more slowly with time and penetration, the clogging advances into the soil and progressively reduces the absorption rate, until a long-term "equilibrium" absorption rate is reached. By "equilibrium" I mean something like this: it might take 1 year to decrease the absorption rate by 10%, 10 years to decrease it by an additional 10%, 100 years to decrease it by an additional 10%, and so on.

Just as for leachlines, there might be a long-term absorption rate for pits. For shallow pits, such rates should be very similar to those for leachlines, which range from 0.8 (in coarse-textured, 1 mpi soils) to 0.2 gallons/square foot/day (in fine-textured soils, with perk times of about 120 mpi).[2,3]

I checked a pit with a diameter of 5 feet and a depth of 20 feet in a very coarse-textured soil (1 to 4 mpi). Although it had been failing for a while, after 20 years of use it was still absorbing roughly 0.3 to 0.6 gallons/square foot/day. Even more interesting are the findings of M. Fernandez (personal communication) and his father. In 1963, they installed 30-foot-deep pits in the Beaumont area, in Riverside County, California. The soil was unstratified clay loam. In 1985, they checked the first of the pits, which had been installed in series. It had been "loaded to the top" for 22 years, but was still absorbing 0.4 gallons/square foot/day. Also, they drilled a vertical hole 4 feet from the pit, and at a depth of 25 feet they encountered black, clogged soil. (So, clogging indeed penetrates deeper than in a leachline.) The pit was full of crusted scum and sludge at the top.

In the Bloomington area, in San Bernardino County, California, the subsurface soils generally consist of strata of sands and gravelly sands (with perk times of about 2 to 4 mpi). Three-bedroom houses are served by pits 5 feet in diameter and 13–16 feet deep (below inlet). The discharge rate in nearby sewered areas of similar socioeconomic status is 280 gallons per average house. On the average, pits last about 14 to 18 years, at a probable gross average loading rate of about 1.3 gallons/square foot/day. (Please note that this rate is obtained from the ratio of 280 gallons divided by 220 square feet of total pit sidewall absorption area. But, actually, most of the time, only a portion of the absorption area is loaded. The first year of operation, maybe only the bottom third of the pit absorbs sewage, and at a rate of perhaps 5 gallons/square foot/day.) As a comparison, on the same type of coarse-textured soils and under the same average loading (280 gallons/day), old "per code" leachlines 50 feet long, 3 feet wide, with 1 to 1.5 feet of gravel below the pipe, have lasted about 8 years in the Upland area, in San Bernardino County, California.

(The loading rates in these leachlines are about 1.1 gallons/square foot/day, counting all of the absorption surface. So the EPA recommendation of 1.2 gallons/square foot/day[2] seems to have fallen short of ensuring long life. Perhaps a large proportion of the leachlines experienced loads much higher than average during some seasons or years. Perhaps the reason for the short life is that the flows were not dosed four times per day, as per Bouma's original recommendation [the EPA recommendation is patterned on Bouma's but it left his dosing schedule out of its Table 7-2].[2] Perhaps Winneberger's more conservative loading rate is more "correct." It is difficult to say.)

Anyhow, it would seem that these pits on coarse-textured soils require *roughly* half the absorption area of leachlines.

On less permeable soils, the conservative long-term absorption rate for leachlines is about 0.2 gallons/square foot/day. It seems plausible that a figure somewhat higher, like the 0.4 gallons/square foot/day measured by Fernandez (in the pit in clay loam), is the equivalent rate for seepage pits.

Therefore, for a range of permeable soils, seepage pits 15 to 30 feet deep might require roughly about half the absorption area of leachlines.

Winneberger mentions that Ryon presented a table giving loading rates for cesspools and seepage pits which were about 69% those of leachlines.[3] (But he noted that Ryon did not discard the bottom of the pits, and the leachlines were 1 foot wide and supposedly had 1 square foot of absorption area per lineal foot.)

If there is such a thing as a factor to convert leachline area to pit (up to 30 feet deep) area, in uniform soils it probably has a value of 0.5 to 1, depending on the depth and diameter of the pit and type of soil, among other things.

9.4 SIZING RECOMMENDATIONS

From all of the above, my recommendations for sizing seepage pits are based on the following premise:

Too many variables influence longevity of pits and not enough is known to predict exact sizes. The prudent policy is to "guesstimate" any reasonable size but to *reserve* enough area on a lot to allow installation of replacement pits to absorb effluent at a long-term absorption rate designed for leachlines. After the first pit(s) fail, superb longevity data for designing replacement pits can be obtained from the failed pit(s). For a given locality with uniform soils, data on neighbors' pit size and longevity can be very useful.

There are many ways to "guesstimate" pit sizes. Size "guesstimates" based on leachline long-term absorption rates may yield pits two or three times larger than those people are used to. It may be impractical and politically unwise to depart drastically from traditional sizes. So pits could be downsized to 1/2 or 1/3, but the reserve area should be kept at full size. Depending on the locality or conditions, I prefer the following sizing strategies:

a. *Coarse-textured strata (alluvial or colluvial gravelly sands to loamy sands, not cemented).* Leachline test perk time, about 1 to 10 mpi; falling head pit test perk time, about 4 to 10 "pit mpi." For such soils, Winneberger's leachline long-term absorption rate goes from 0.8 to 0.4 gallons (of sewage)/ square foot/day;[3] and EPA's, from 1.2 to 0.8 gallons/square foot/day.[2] Now, a gross application rate of 1.3 gallons per square foot per day should yield a useful life similar to that of the Bloomington pits (see Section 9.3), which is not long enough. A gross application rate of 1 gallon/square foot/day (which is in the ballpark of EPA recommendations) should increase longevity 1.3 times due to increase in area, plus up to (but not quite) another 1.3 times due to the increase in depth of the pit (and concomitant increase in surface head). Thus, the longevity would increase to roughly 21 to 27 years. (Note that this increase might not be cost-effective if the additional depth requires the use of expensive materials, equipment, and methods, or if sewering is anticipated.) Few perk tests, if any, are needed in these coarse-textured materials; often a visual inspection of the strata is sufficient.

b. *As above, but some lower-permeability strata are evident in the soil profile.* Proceed as above, but disregard the lower-permeability strata in the computation of absorption surface and depth.

c. *The strata are too mixed up to tell what the soil profile is like, or the strata are too numerous to test individually.* Use the falling head formula. If the test results are equal to or lower than 10 pit mpi, no problem; assume the strata are quite coarse-textured and proceed as in "a" or "b" above. If the test results fall between 10 and 29 pit mpi, use a "guesstimated" disposal rate of 0.4 to 1, say 0.7 gallons of sewage/square foot/day. If the test results fall between 30 and 60 "pit mpi," the 0.4 gallon rate might or might not be too conservative.

If they exceed 100, the soil strata might have poor permeability or the test hole sidewalls might be highly compacted. Such results may be subject to considerable error. (Above 100 pit mpi, I would try not to use pits: the costs associated with frequent pit replacement or with running out of replacement area can be substantial.)

(Caution: In some soils that might have permeable upper strata and low-permeability lower strata, it might be advisable to do the following in some of the test holes when you use the falling head formula: after the last refill, measure how fast the water level keeps on falling until it gets down to the bottom or slows down considerably. Plot these measurements, look for sudden changes in the *rate of fall*, and see if most of the absorption takes place in just one stratum near the bottom, or near the middle, or near the top of the 20- to 40-foot-deep test hole. Also note that in soils with horizontal stratification, the falling head test indicates how fast the water moves laterally during the short testing time. But the critical and unknown parameter is how fast water moves through the most-limiting layer below the test hole after steady state is reached. See Chapter 13.)

9.5 PROBLEMS

9.5.1 Derivation of the Falling Head Formula

Derive the formula, in gallons (of water absorbed) per square foot per (extrapolated 24-hour) day, and in "pit mpi."

Let Q = volume of water absorbed during a measurement interval = $\pi r^2 F/t$ (ft³/hour).

Let S = amount of sidewall through which absorption takes place = $2\pi rL$ (ft²).

Hence, $Q/S = rF/2Lt = DF/4Lt$ [(ft³/hour)/ft²] where D = diameter of test hole

Multiplying DF/4Lt by 7.5 (gallons/ft³) and then by 24 (hours/day), we obtain

$$Q/S = 45FD/Lt \text{ (gallons/ft}^2\text{/day)}$$

If we were to assume, wrongly, that sewage is absorbed five times more slowly than water, we would divide the right-hand side of the formula by 5 and obtain the currently popular formula, $Q = 9FD/Lt$ (gallons of sewage/ft²/day). Note that in this formula the S is omitted but assumed.

Now, 1 (gallon/ft²/day) × (1/7.5) (ft³/gallon) = 0.133 (ft/day). And 0.133 (ft/day) × 12 (in./ft) = 1.6 (in./day). And 1,440 (minutes/day) over 1.6 (in./day) = 900 mpi. Hence, mpi = (900/45)FD/Lt = 20FD/Lt.

9.5.2 Effect of Surface Head

When testing in uniform coarse-textured soils, one may obtain very similar stabilized Q's (or mpi's) whether the water column (surface head, or "L") is, say, 5 or 20 feet. Explain why. (Hint: review Chapters 5 and 6.) The explanation follows.

About 1978, a new perk tester, Sandy M., deviated from the standard procedure. The Q's have to be measured with an L or h equal to the proposed depth of the seepage pits. But she used surface heads of 5 and 20 feet when testing a uniform, coarse-textured soil, and measured the same Q's in all cases. After I evaluated her report, I scolded her: "We don't know what these results mean." At that time, I did not believe that absorption could be directly proportional to the surface head, and that the Q's could be the same in spite of surface head differences, as her results showed. And the literature, then as now, doesn't either.[3,4] So I assumed her results were just a fluke. (Sorry, Sandy, I goofed.)

Since then, I have seen other instances in which the Q's were the same though the surface heads were not. I have checked published data[5,6] that purport to show that absorp-

round vertical hole

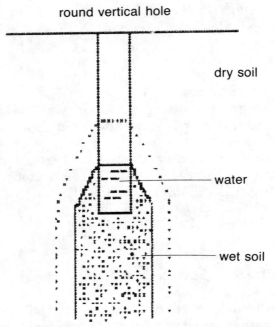

dry soil

water

wet soil

Figure 9.1 Vertical cross section of a round vertical test hole in a uniform soil. Water flows out of the test hole and wets the soil as shown.

tion is *not* proportional to surface head, but I have interpreted the data in a very different way.

Today, as far as I can see, the question is not whether absorption is or is not directly proportional to h. The question is, "When is it and when is it not?"

If we have a situation like that shown in Figure 5.4, in which the wet soil is confined within a vertical tube of infinite depth, we have that, as long as X > > h, a change in h will not affect the absorption rate (see Section 5.1). But when water is not confined within a tube, water flows downward and laterally out of a hole and forms a (truncated) cone of supposedly saturated soil, as shown in Figure 9.1. If we increase h, the cone would grow as shown with the dotted

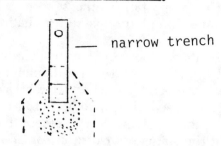

Figure 9.2 Vertical cross section of a trench (narrow leachline) in a uniform soil. Water wets the soil as shown.

line. At first sight, we might assume that the radius at the base of each cone is proportional to h, but this is not so.

If instead of a hole we would have a narrow leachline, the triangular cross sections in Figure 9.2 indeed would have bases proportional to the h's: the gradients impelling the water laterally are proportional to the head or height of the water and inversely proportional to the lateral path. So, if X > > h, and we double h, we will double the base of the triangle; or what's the same, the cross-sectional area of saturated flow will also double, and so will the absorption.

But if we have a round vertical hole instead of a narrow horizontal trench, we cannot expect that the radius at the base of this cone will double when we double h. If this were to happen, the cross-sectional area at the base (or below) this cone would quadruple (since it is proportional to the square of the radius), and so would the absorption or total saturated flow downward. And if we quadruple the h, the base would increase 16 times, and so would the absorption. But this does not happen in reality. The Q/S measured with h = 30 feet is very similar to that with h = 1 foot, and nowhere near 30 times larger (i.e., square of 30, divided by 30). The reason is that it is not the radius that increases in proportion to h, but the basal area of the cone. And the reason for this, in turn, is that the horizontal component of the gradient decreases not in proportion to distance, as in the narrow leachline (or "line source"), but in proportion to the square of the radial distance. (The same thing happens with radiation and concentration gradients.)

In short, we may expect that in an ideal coarse-textured isotropic soil (if X>>h and if the diameter of the hole is negligible with respect to h), when we increase h by a factor of N, we also increase total flow by a factor of N; the absorption per square foot of sidewall will remain the same. On the other hand, if the soil is not isotropic, or if any of the other conditions do not hold, deviations are to be expected.

9.5.3 Pit Costs and Diameters

The EPA Design Manual (in its Table 7–6) refers to pits of up to 12 feet diameter. If the cost of a pit were proportional to the amount of soil excavated, how much more expensive (per square foot of sidewall absorption area) would be a pit twice the diameter of another? How much more would h (or water level) rise the instant we dump the same volume of water into each of two pits, one twice the diameter of the other?

9.5.4 One Sizing Procedure for Pits

In a certain jurisdiction, the way to size a pit is as follows: "Excavate the seepage pit (minimum depth 10 feet below inlet) and line it. Soak. Refill after 24 hours. After another 24 hours come back and calculate water loss. It must be at least 5 times the daily flow, and not less than 0.83 gallons per square foot of absorption area per 24 hours."

I asked the local sanitarians how well this procedure worked. They said they were very satisfied with it, and did not know of any failures. I asked what kind of soils they had. "Very sandy," they replied.

Would you use their procedure? Explain. The answer is in Appendix E.

REFERENCES

1. U.S. Public Health Service. 1967. Manual of Septic Tank Practice. (Pub. No. 526.)
2. EPA Design Manual. 1980. EPA-625/1–80–012.
3. Winneberger, J.T. 1984. *Septic Tank Systems*. Butterworth Pub., Stoneham, MA.
4. Winneberger, J.T., and Klock, J.W. 1973. Current and Recommended Practices for Subsurface Waste Water Disposal Systems in Arizona. ERC-R-73014. College of Engineering and Applied Sciences, Arizona State University, Tempe, AZ.
5. Winneberger, *Septic Tank Systems*, pp. 43 and 53.
6. Winneberger and Klock, Current and Recommended Practices, p. 64.

10

Various Onsite Sewage Disposal Technologies

10.1 "INNOVATIVE" TECHNOLOGIES

When I began working as a septic systems specialist in San Bernardino County, I heard about past problems with "innovative" onsite sewage disposal technologies:

a. Pumping of sewage to uphill locations was not permitted for new construction, because users could not or would not maintain the pump system. Now, all flows had to be gravity flows.

b. Aerobic systems had malfunctioned, because users were not interested in maintenance and repair. (Also, some people were bothered by the pump noise, and "pulled out the plug.") Although they could be installed, they had to meet septic system standards and their leachfields were sized just as for septic systems.

c. Spraying pressurized clarified sewage over a large area with thin soils did not work very well. The sprinkler heads occasionally plugged up, and they froze in winter.

d. In the case of incinerating toilets, the toilet representative or repairman often lived far away and one had to wait weeks for replacement parts.

In the ensuing years, I saw fit not to approve a variety of "innovative technology" proposals. Two of these, I can't forget.

One was promoted by an elderly gentleman who said he

represented a large, well-established eastern company, "B."
He wanted to try (for the first time) to discharge aerobically
treated sewage effluent through drip-irrigation tubing. A
developer of a large residential project was "sold" on it. I
suspected that if drip-irrigation had problems with plugged
emitters (which discharge clear irrigation water into the soil),
it would have problems discharging sewage. I called the
alleged parent company, "B." They told me that they knew
the subject gentleman, and that they had nothing to do with
him or his projects. My supervisors politely informed him
that he would have to prove his system worked before it
could be approved.

The gentleman vanished. But a few months later I received
a call from a central California county sanitarian. He wanted
to check up on the performance of an aerobic sewage drip
system an elderly gentleman (of familiar name and description) claimed to have installed in San Bernardino county; the
following day this gentleman was going to speak in front of
the sanitarian's county board of supervisors.

As for another innovative system, its promoter maintained
emphatically that God had shown him how to design it. But
as yet he has not brought to me the hard data I requested. (In
another county, this system turned out to be a fiasco.)

I was favorably disposed to other types of proposals, but
for various reasons they did not "take root."

An evapotranspiration (ET) system is a large basin with
impermeable walls filled with gravel and sand and topped
with a layer of soil. Effluent is discharged into the gravel,
taken up by plants growing on the layer of soil, and lost by
evapotranspiration.

Two ET proposals I evaluated were not backed up with
reliable data, were extremely expensive, and offered absolutely no advantages over conventional septic systems at the
proposed sites.

It is not difficult to follow guidelines such as Ingham's[1] and
come up with an engineered ET proposal; but it is quite
difficult to account for the ET system's implied and explicit
constraints. Salts may accumulate, plants may die, weather

might not cooperate, onsite microclimate might differ from areawide weather used for design, users may not maintain, etc.

When there is little soil above impermeable bedrock or groundwater, or the soil is excessively permeable or impermeable, a specially designed mound of soil-covered sand is placed upon the ground and used as a filter to purify and spread the effluent. It is called a "mound system." The effluent is dischaged under pressure through perforated pipes in the sand. Details are given elsewhere.[2] When I was consulted regarding the use of mounds in thin, moderately sloping soils over impermeable bedrock, I noticed that, in the prospective sites, it would cost half as much to raise the ground level (with fill available at the site) and to install conventional leachlines instead of mounds. (The cost of an engineered mound was $10,000.) Tips on fills are given in Section 10.3.

On many occasions I was asked to allow installation of composting toilets. After reading the claims made in the proprietary literature, I thought highly of those devices. However, after I found out that the systems required much conscientious, elaborate, and frequent maintenance work by the user, I became suspicious. I enquired further, and eventually found out that indeed composting toilets were considered to be a potential public health hazard, as viable parasites and pathogens had been found in the compost.[3] In 1983, the "Liquid-Waste Committee" of the California Conference of Environmental Health Directors recommended that composting toilets not be allowed even for experimental purposes.

Years ago I witnessed a demonstration of the patented electro-osmosis (E-O) system. I saw that the water level was practically stationary in perk test holes dug into a clay (perk time, about 150–240 mpi); yet when the E-O electrodes were installed, the water level fell rapidly (about 10 mpi). More recently, two investigators found that E-O might not make any difference in the long run in "average" soils; the long-term absorption of sewage (rather than water) by leachlines in a 7 to 8 mpi soil was governed by the lower long-term absorption rate after a clogging mat formed.[4]

Over the years, I have become a firm believer in Murphy's law—"If anything can go wrong, it will." The more complicated the "innovative" system, the higher the likelihood that it will break down and cause problems. And if it is proprietary or exclusive, it might be difficult to obtain parts or service.

In order to promote innovation, I have encouraged anyone who has a reasonable idea to try it out at his/her own risk, by means of a public notice. (Details are in Appendix F.) But as yet I have seen nothing that can equal the cost-effectiveness of a septic tank with leachlines, seepage pits, or deep leachlines (with 5 to 10 feet of gravel depth below the perforated pipe.) These deep leachlines are not specifically authorized by the Uniform Plumbing code, but may be used effectively under proper soil and "political" conditions.

10.2 COMPARISONS AND RECOMMENDATIONS

The leachline has the following advantages over the seepage pit:

a. It may be installed where groundwater is or might be relatively shallow.
b. Its absorption area requirements are fairly well-defined in the literature.
c. It is easy to install with a backhoe or even with hand labor.
d. If it becomes clogged and "fails," a backhoe operator can scrape a couple of inches off the bottom and sides (under dry soil conditions), and rebuild the leachline. (This cannot be done with deep seepage pits.)
e. It is safe. No one can fall into an old abandoned leachline. (I have heard of people falling into abandoned pits.)
f. It may accomplish a higher degree of sewage aerobic biodegradation than a pit.
g. It may cost less to install.

The advantages of the seepage pit over the leachline are:

a. It requires less (horizontal) area on a lot.
b. It can reach permeable strata at great depth.
c. It is little affected by rain or lawn watering, by roots, and by traffic over it.
d. High surface head and sidewall, and larger fluctuations in height of effluent column (per volume of effluent), tend to increase useful life.

I have checked claims and information provided by many "innovative" onsite system manufacturers. Invariably, the claims turned out to be highly exaggerated, and the "information" without solid foundation.

Again, as far as I can see, the simplest and most cost-effective onsite disposal technologies depend on nothing more than: a septic tank, a seepage pit or leachlines, properly placed fills, and drop boxes. No distribution boxes.

Per code requirements (Uniform Plumbing Code), a distribution box supposedly splits the sewage flow evenly into 2, 3, or more portions and directs them to an equal number of pits or leachlines. But I have observed that a septic tank buffers the flow of sewage, so that flow trickles into the distribution box, and goes preferentially into one of its outlets and shuns the others. Hence, one might find a failing leachline beside a dry leachline, though both are served by the same distribution box. This poor performance has been reported in the literature.[5,6]

One solution is to install an automatic siphon between the leachfield and the septic tank, so as to discharge sewage in massive batches. Massive flows have a chance of being split evenly by a distribution box. But siphons do malfunction and require maintenance.[7]

In sloping ground, the best strategy is to eliminate distribution boxes, and to allow the sewage overflows to cascade sequentially from higher to lower leachfields by means of drop boxes, inverts, or step-downs. In flat ground, the ends of parallel leachlines fed by the same distribution box should be connected with a perpendicular graveled trench, so that the sewage flow can move from one leachline back into the other.

Regarding simple and inexpensive technical devices, the most promising one I have seen in the last few years is the "dipper." The "dipper" is a plastic tray on a pivot. The tray receives trickling sewage from the septic tank and retains 1.5 gallons. When full it dips and discharges the 1.5 gallons in less than 2 seconds. This massive flow goes to a distribution box, where it is properly split. The problem of non-uniform distribution of sewage flow by distribution boxes might be at an end if the "dipper" performs as claimed. Information about this device can be obtained from United Concrete Products, 173 Church St., Yalesville, CT 06492, (213)269–3119.

10.3 FILLS FOR LEACHLINES

As a medium for wastewater disposal, a good fill can be as good or better than many a natural soil. The safest way to construct a good fill follows:

1. Place a layer of dry loamy sand about 8 inches thick over the dry ground that is to be built up.
2. Smooth out and press down the layer of dry loamy sand with a bulldozer or scraper.
3. Repeat the procedure with successive layers until the ground has been raised about 7% beyond design height.
4. After at least one wet and one dry season have given the fill a chance to settle, check the perk rate, remove or add up to a foot of fill if necessary, and then install leachlines. (If less than two feet of fill have been placed, it may not be necesary to wait for the fill to settle.)

Attention to detail and some expertise are called for when deviating from the procedure described above.

Let's assume the fill materials are coarser-textured. Then, the big soil pores may not filter out sewage microbes.

Let's assume the fill materials are finer-textured. If they are dry and lumpy when placed in layers, the resulting fill may have big pores or cavities and may not filter out sewage microbes very well. But if the fill materials are barely moist (just to the point where a change from the dry state is noticeable), they may smear and compact rather easily, especially so just below (bulldozer) tracks or tires. The result is that a vertical cross-section of the fill may show compacted layers of soil stacked one upon another, with an even more compacted or smeared area marking the transition between adjacent layers. This condition is common in foundation fills (which must achieve at least 90% of maximum compaction) compacted with tamper, bulldozer, heavy equipment tires, or steamroller. (A sheepsfoot roller may or may not result in clearly displayed layers.)

If one must use slightly moist, finer-textured fill materials, one should do at least the following: disk the upper 2 or 3 inches of each layer before another layer is placed on top; trench and observe the profile of the fill for signs of highly compacted or smudged zones between layers; do a perk test; after completing this test, excavate around the perk test hole and see if the wetting front spread laterally (sideways) along fill layers and failed to move downward. (If this is the case, the fill might not be usable).

Cobbles and stones may compact adjacent fill materials and leave big holes between themselves and the fill material.

REFERENCES

1. Ingham, A. 1980. Guidelines for Evapotranspiration Systems. State Water Resources Control Board, P.O. Box 100, Sacramento, CA 95801.

2. Ingham, A. 1980. Guidelines for Mound Systems. State Water Resources Control Board, P.O. Box 100, Sacramento, CA 95801.
3. M. Smith. 1981. Evaluation of Compost Toilets. USDA Forest Service Equipment Development Center, San Dimas, CA 91773; and unpublished data from the California Health Department.
4. Effert, D., and Beer, C. In:ASAE Pub. 07–85. 1985. Proceedings of the Fourth National Symposium on Individual and Small Community Sewage Systems, pp. 59–68, ASAE, St. Joseph, MI 49085–9659.
5. Mellen, W. In:ASAE Pub. 07–85. 1985. Proceedings of the Fourth National Symposium on Individual and Small Community Sewage Systems, pp. 87–93, ASAE, St. Joseph, MI 49085–9659.
6. Gross, M., and Thrasher, D. In:ASAE Pub. 07–85. 1985. Proceedings of the Fourth National Symposium on Individual and Small Community Sewage Systems, pp. 260–264, ASAE, St. Joseph, MI 49085–9659.
7. Converse, J. C., et al. In:ASAE Pub. 07–85. 1985. Proceedings of the Fourth National Symposium on Individual and Small Community Sewage Systems, pp. 95–103, ASAE, St. Joseph, MI 49085–9659.

11

Degradation of Groundwater by Septic Systems

Septic tank effluent discharged into soil carries various constituents which may reach and degrade groundwater. Some may be natural chemicals, such as chloride, nitrate, and phosphate salts. Some may be chemicals such as waste oil fractions, fuel oil, TCE and other chlorinated hydrocarbons, gasoline, turpentine, and almost anything that can be found in a house and flushed down the toilet. Other constituents are parasites and microbes of various kinds.

In the period of 1971 to 1982, waters from wells tapping into degraded groundwater have accounted for at least 177 disease outbreaks involving 30,046 people nationwide.[1] Most of these people were affected by gastroenteritis and shigellosis. (Most of the gastroenteritis appears to be due to viruses.) Twelve of the outbreaks were caused by chemicals (petroleum products, nitrate, PCBs, TCE, phenols, benzene, arsenic, selenium) and involved 157 people.[1]

Let us start with the problem of microbial pollution.

11.1 SOILS AS A SIEVE

When they receive septic tank effluent, soils act like sieves and easily trap large parasites like flatworms, roundworms, and tapeworms. The ova of these parasites are generally fairly large, 50–60 μ (microns or micrometers) in diameter for *A. lumbricoides*. The diameter of the smallest cyst of the protozoan *Giardia* is 7 μ.[2] Leachfields capable of trapping intestinal bacteria will also trap larger organisms, so we need not give further attention to these larger organisms. According to a chart prepared by Osmonics, Inc. (1984), the diameters of bacteria range from about 30 to 0.25 μ, and of viruses, from about 0.1 to 0.009 μ. As the microbes get smaller than about 3 μ, physical trapping becomes less efficient, but other factors play an increasingly more important role in retaining microbes within the soil.

11.2 MICROBIAL RETENTION WITHIN THE SOIL

Research publications regarding individual measurements of microbial travel within soils seem to be inconsistent. Some report that bacteria (or viruses) have traveled hundreds of feet within the soil; others, that they have been retained in less than a foot of soil. Two main reasons for this apparent inconsistency are:

a. The conditions in which the travel or retention took place were different; all of the variables affecting travel or retention were not controlled or accounted for.
b. Every specific bacteria (or virus) is different from every other specific bacteria (or virus). For instance, one particular type of virus may be much more adsorbed by a given soil clay at a given soil pH than another type of virus. One type of bacteria may survive longer and travel farther in a viable state than another type.

Of all the articles I have read on the subject matter, three[3-5] shed more light than all others, and are expounded upon below.

Hagedorn et al.[3] reviewed pertinent literature and arrived at the following conclusions (slightly modified from the original text):

1. Microorganisms move only a few feet in unsaturated soil, but much larger distances in saturated soil.
2. Bacterial retention is higher in finer-textured soils.
3. The main limitation to travel through soil is physical straining or filtration (of bacteria or larger microbes).
4. Adsorption plays a role in retention of bacteria, and increases with clay content.
5. Death of the microorganism plays an important role. Death may occur due to ingestion by other organisms, adverse soil conditions (no nutrients, drying, antagonistic organisms' secretions, such as antibiotics), and "aging" during long retention periods.

Hagedorn et al.[3] reviewed experimental results published by Bouma in 1972. These indicated that 1 to 3 feet of unsaturated soil below a leachline's clogging mat was adequate for complete bacteria removal. So, Hagedorn et al. concluded that the (USPHS Manual 526) standard of 4 to 5 feet of "suitable" soil fell in line with Bouma's experimental data.[3]

However, the EPA Manual's[6] standard minimum separation between leachline bottom and groundwater is 2 to 4 feet. I think that this is *inadequate*:

a. Much of the knowledge about wastewater purification below leachlines is derived from studies about distribution of bacteria in the soil profile after a clogging mat develops. Bacterial and especially viral movement can be more extensive before the mat is fully developed.
b. Capillary "suction" can maintain the continuity of large-diameter soil water columns from the bottom of the leachline to groundwater, and decrease "sieving" or filtration.
c. Coarse-textured soils absorb much rainfall; absorbed rainfall creates a layer (or layers) of nearly saturated soil. While the soil is draining, bacteria and viruses may be "washed down."
d. Mounding of effluent can raise the level of groundwater. (This will be addressed in Chapter 13.)
e. As pointed out in the previous chapter, minimum requirements become the standards of practice. Soil conditions throughout the land are quite more varied than those near

a few experimental leachlines. Few perk professionals know when (or dare) to exceed the minimum requirements.

f. Depth to shallow groundwater is at least as variable as the weather, and usually cannot be predicted with certainty. (It can be predicted if drains are installed.) As for using iron bars or looking for mottles, such techniques are not reliable, in light of the variables involved (time of submergence, rainfall, presence and location of decomposable organic matter, parent material, etc.) For instance, Couto et al.[6] found no indications of groundwater in a soil they knew had groundwater at 2 feet for more than 120 days per year. A good 5 feet may promote more complete decomposition of biodegradable (possibly mutagenic) organic matter. This is purely speculative, but considering that sewage shows mutagenicity per the Ames test, and that algae have been found at 5 feet below ground living only on dissolved nutrients coming from above, it might be advisable to maximize opportunities for biodegradation by requiring at least 5 feet to groundwater, even if 2 or 4 feet were sufficient to trap bacteria.

g. Not enough data exist to ascertain a safe distance of travel through soil for trapping all types of viruses.

Movement of typhoid bacteria from a leachline down through at least 3 or 4 feet of "less than 1 or 2 mpi sand," and thence through 210 feet of saturated soil (i.e., groundwater) downgradient to a well was reported by McGinnis and De Walle.[4] These authors also tabulated the distances traveled by other types of bacteria (as reported in other publications); distances ranged from 1 to 300 feet in sandy aquifers, reached 2,800 feet in gravelly aquifers, and 3,300 feet in a fractured limestone. They pointed out that the (code-required) exclusion of leachlines from an area 100 feet around a drinking water well does not ensure adequate protection, particularly if the well is downgradient and the velocity of flow is high.

What is true of the extent of bacterial movement is just as true of viral movement in soils and groundwater. Vaughn and Landry reviewed 182 publications and put together a comprehensive picture of viral movement.[5] The next two paragraphs contain abstracts from their paper.

Regarding travel, they referred to a report of poliovirus moving downward from a leachfield and then laterally 300 feet (91.5 m) to a well. Under rapid-infiltration units receiving secondary-treated sewage, viruses were recovered 30 m down and 183 m lateral distance. One report concerns Coxsackie virus 402 m downgradient from a sanitary landfill; another report concerns a coliphage which moved 900 m under saturated soil conditions at a rate of 350 m/day.

When they enter a leachfield, viruses may be metabolized (or "eaten") by microbes, or they may adsorb onto the slimy bacterial secretions in and around the clogging mat, or they may continue their travel and be adsorbed in charged soil particles (mostly clays), and inactivated by (the fairly common clay fraction constituents) Al_2O_3 or MnO_2. Adsorption to the (usually abundant) SiO_2 and Fe_2O_3 does not inactivate viruses. Rainfall can desorb viruses and carry them down in a still viable state. Virus adsorption is extremely variable: different strains of the same virus may adsorb to a different extent; and within a purified population of virus particles, some subgroups adsorb at different rates. Dry soil appears to kill or inactivate viruses.

To make a long story short, suffice it to say that the old concepts about separations between leachfields and groundwater and wells need updating.

Standards currently in vogue seem to have been influenced by a single study published in 1970, in which Romero cataloged and interpreted the data then known on microbial travel through soils.[8] Today, it is recognized that not enough is known to predict how far each specific virus can travel through each specific soil,[5] particulary under actual field conditions (mass loadings, rainfall-saturated soil, not fully developed clogging mats, and root channels). However, it can be safely said that the control of viral pollution depends mainly on forcing sewage effluent to pass through a biomat and/or through unsaturated soil with sufficient adsorption sites within its matrix (mostly from clay or silt). Other things being equal, on a per-foot-of-travel basis, comparatively little viral adsorption and inactivation occur after viruses reach groundwater and move through saturated soil. Generally, the vertical separations between leachfields and groundwater are far more important than the horizontal separations between leachfields and wells.

11.3 VERTICAL SEPARATIONS BETWEEN LEACHFIELDS AND GROUNDWATER OR IMPERMEABLE BEDROCK

In order to promote unsaturated flow conditions below a leachfield, minimum separations from the bottom of leachfields to groundwater or impermeable bedrock (or impermeable stratum) have been adopted in various quarters. Unsaturated flow may enhance the activity of aerobic soil microbes which "eat" pathogens; also, unsaturated flow moves in thinner films and through smaller-diameter pores, thus enhancing the sieving and adsorption of pathogens. Machmeier's recommendations follow.[8]

> Use a minimum separation of 3 feet.
> If soil "A" perk time is less than 0.1 mpi, little filtration of pathogens will occur. Place 6 inches of 6 to 15 mpi soil between soil "A" and leachline gravel.
> If a soil's perk time is between 0.1 and 5 mpi, bacteria will move through with the percolating sewage until a biomat forms and filters the bacteria out. Leachlines can be installed in such soil but the leachline must be designed so as to promote the formation of a biomat; for instance, by dividing the total length of the leachline into 4 parts in such a way that the overflow from the first part goes to the second, and so on sequentially. Or else, leachlines can be installed but the sewage is spread uniformly over the whole leachline under pressure distribution. (Spreading the sewage over a large surface results in unsaturated flow.) Or else, 6 inches of finer-

textured 5 to 15 mpi soil is placed as explained in the paragraph above.

The EPA Design Manual[7] recommends 2 to 4 feet of vertical separation, and a different approach: If the soil perk time is less than 1 mpi, place at least 2 feet of "loamy sand or sand-textured soil" between the soil and the gravel of the leach-line. (There is no definition of sand-textured soil within the vocabulary of soil science; however, the EPA Manual implies in a subsequent paragraph that this is 6 to 15 mpi soil.)

According to the Manual, if perk time is higher than 1 mpi, conventional leachlines can be installed. (This doesn't seem right, in view of the paragraph above, where at least 2 feet of 6 to 15 mpi soil are required. The explanation for this discrepancy may be found in Winneberger's book.)[10]

According to Winneberger,[10] the origin of the 1 mpi standard is more or less as follows:

In 1978 a conference was held in EPA offices . . . 5 mpi was considered as a limit . . . One consultant pointed out that such limit had no technical merit and would likely rule out disposal fields in Florida, where soils had lower mpi . . . One of the less experienced conferees suggested 1 mpi: "How about that?" . . . And the motion carried!

There is nothing magical about a 1 or a 5 mpi standard, or about a 2 to 4 or a 5 foot separation. However, it can be said that "average" soils with no channels or fractures are generally safe if the perk time is higher than 5 mpi and the sewage percolates at least 5 feet through unsaturated soil. This rule— "5 mpi and 5 feet"—is not absolutely foolproof, but it is a convenient starting point for evaluation of site suitability. A "2 and 2" or a "1 and 2" rule might work sometimes, but it would imply considerably more risk, and would involve extreme care in the planning, construction, use, and perhaps monitoring of systems and groundwater. I doubt that most jurisdictions would feel comfortable with such impositions, unless they are prepared either to treat local water pumped for domestic consumption (by coagulation, filtration, and chlorination or ozonization), or else to avoid using such water for domestic consumption. Finally, we should note that if pits are installed in series, the first pit may receive a tremendous amount of flow under a high surface head, and saturated flow may occur under it for more than 10 feet downward, especially if the clogging mat has not had time to form. Pits should not be installed in series where groundwater may be at risk.

11.4 CONTAMINATION BY EXTRANEOUS CHEMICALS

Nowadays we are well aware that a variety of carcinogenic synthetic chemicals have contaminated groundwater in many areas of the United States. On or about March 1986, CBS's "60 Minutes" presented a segment about children who died of leukemia because a chemical company dumped a chlorinated hydrocarbon solvent or degreaser. The chemical found its way to the children's neighborhood water supply. The owner of the company was sued in federal court. The parents were compensated per the undisclosed provisions of an out-of-court settlement.

As recently as 1979, things were different. The *Cape Cod Times* (September 3, 1979) reported that local boards of health were reluctant to take action against the sale of chemicals used as drain and cesspool cleaners that were polluting groundwater supplies. The reason for the reluctance was that a company making and selling such products had sued a county health department for libel ($15 million), for implying that their drain cleaner (with methylene chloride and trichloroethane) was hazardous to people drinking groundwater contaminated by it.

Today any such lawsuit would be thrown out of court, laughingly. But that's all that has changed. The federal government still does not control or intend to control the improper use and disposal of small quantities of hazardous chemicals which can be purchased in most supermarkets.[11,12] The responsibility to do so falls on state and local jurisdictions.[12] How well they deal with this issue is "a measure of their strength, capabilities, and resources."[12] (Ah, diplomacy.)

In California, we have nine Water Quality Control Board Regions. Their mission is to protect the waters of the state.

A while ago I found a septic system declogger with characteristic odors for sale in a local supermarket, and I notified the local (Santa Ana) Regional Board. It made the manufacturer change the formula of the declogger very quickly.

The manufacturer of another declogger openly stated that it contained chlorobenzene, in an ad placed in a national environmental health magazine (of all places)! When I wrote and questioned the legality of their product, the manufacturer's legal counsel showed me a letter from an EPA officer stating that it was okay to use it: It was not regulated under the Resource Conservation and Recovery Act because it was used for something (declogging) rather than dumped. I publicized this experience,[11] and informed the (Santa Ana) Regional Board of what was going on. Its executive director wrote a terse note to the manufacturer's legal counsel informing him that groundwater was an important local resource, and that under no circumstances would he permit the use of the declogger. I never saw the ad again in any magazine.

The problem is still with us. Few jurisdictions have a comprehensive program to educate septic system users to the fact that they may end up drinking many of the chemicals they flush down. Even fewer control the products that are sold for sewer or septic system "declogging." As far as is known, elves are reluctant to check supermarket shelves for decloggers and to educate septic system users. (Hopefully, sanitarians do.)

11.5 DEGRADATION BY NATURAL SEWAGE COMPONENTS

Under "average" conditions, after sewage percolates through 5 feet of unsaturated soil it should have negligible amounts of BOD and SS; respectively, less than 2 and 1 mg/L.[13] Sewage TDS (total dissolved solids)—also called TFS (total filterable solids)—corresponds mainly to salts. Precise definition of this (and other terms like BOD, SS, etc.) can be found elsewhere.[14] Most of the salts in piped-in water are chlorides, sulfates, and bicarbonates of sodium, calcium, magnesium, and potassium. TDS may change a little in composition when it goes through the soil. Some cations may exchange with those in the "soil exchange" colloids, or some TDS salts may precipitate, or soil salts may be dissolved. TDS is definitely not "absorbed" by the soil or the leachfield. TDS moves in the percolating liquid and may reach groundwater and increase its salinity. A rule of thumb is that water which passes through a septic tank experiences a TDS increase of about 100 to 300 mg/L.

The phosphorus (mainly phosphate salts) found in septic tank effluent is not directly harmful to humans. But if it reaches lakes or bodies of water it may act as a "fertilizer" and promote excessive growth of algae (eutrophication). (When algae accumulate and decompose, the water body assumes a disagreeable aspect and odor; fish may die.) Phosphate anions are precipitated by calcium, iron, and aluminum cations, one or more of which are quite abundant in most soils; and the phosphate anions adsorb to soil sesquioxides and calcareous precipitates. So, generally speaking, phosphorus is not much of a problem. It may be a problem if leachfields are located in some very coarse-textured (less than 1 mpi) soils near water bodies.

For evaluation purposes, the EPA assumes that each person on septic systems within 300 feet of a lake shoreline releases 0.25 pounds of phosphorus (5 to 10% of the total yearly per capita output) to lake waters.[15] Sewage effluent contains about 10–14 mg/L of phosphorus,[16] while natural water bodies contain 0.002 to 0.09 mg/L.[17]

Canter and Knox[16] refer to a study conducted by Viraraghavan and Warnock in a clayey Canadian soil with cold, snowy winters and wet springs (melting snow). In this study, (apparently) 5 feet of soil reduced septic tank effluent SS, BOD, and COD by an average of 75 to 90%; this reduction was higher in summer and fall (when the soil was more unsaturated), and lower in winter and spring. The reduction in phosphorus was only 25 to 50%, much lower than that reported in the literature. (Possibly, the cold temperature and saturated soil allowed a high proportion of organic phosphorus to pass through without precipitating.)

Nitrates are another pollutant of concern where septic systems are used. They are discussed in the following chapter.

REFERENCES

1. Craun, G.F. 1985. A summary of waterborne illness transmitted through contaminated groundwater. *J. Environ. Health* 48: 122–127.
2. Logsdon, G., et al. 1979. Water filtration techniques for removal of cysts and cyst models. In Waterborne Transmission of Giardiasis, U.S. EPA 600/9–79–001.
3. Hagedorn, C., et al. 1981. The potential for groundwater contamination from septic effluents. *J. Environ. Qual.* 10: 1–8.
4. McGinnis, J., and De Walle, F. 1983. The movement of typhoid organisms in saturated, permeable soil. *J. Am. Water Works Assoc.* 75: 266–271.
5. Vaughn, J.M., and Landry, E.F. 1983. Viruses in soils and groundwaters. In *Viral Pollution of the Environment*, pp. 163–210, G. Berg, Ed. CRC Press, Boca Raton, FL.
6. Couto, W., et al. 1985. Factors affecting oxidation-reduction processes in an oxisol with a seasonal water table. *Soil Sci. Soc. Am. J.* 49: 1245–1248.

7. U.S. EPA. 1980. Design Manual: Onsite Wastewater Treatment and Disposal Systems. EPA-625/1–80–012.

8. Romero, J.C. 1970. The movement of bacteria and viruses through porous media. *Ground Water* 8: 37–48.

9. Machmeier, R. 1981. Town and Country Sewage Treatment. Extension Bulletin 304, Agricultural Extension Service, University of Minnesota, St. Paul, MN 55108.

10. Winneberger, J.T. 1984. *Septic Tank Systems*. Butterworth Publishers, Stoneham, MA.

11. Kaplan, O.B. 1983. Some additives to septic tank systems may poison groundwater. *J. Environ. Health* 45: 259.

12. Bacon, J.M., and Oleckno, W.A. 1985. Groundwater contamination. *J. Environ. Health* 48: 116–121.

13. U.S. EPA et al. 1981. Process Design Manual For Land Treatment of Municipal Wastewater. EPA 625/1–81–013.

14. U.S. EPA. 1974. Methods for Chemical Analysis of Water and Wastes. EPA 625/6–74–003.

15. U.S. EPA Region V. 1983. Final-Generic EIS: Wastewater Management in Rural Lake Areas. Water Division, Chicago, IL 60604.

16. Canter, L., and Knox, R. 1985. *Septic Tank System Effects on Ground Water Quality*. Lewis Publishers, Chelsea, MI 48118.

17. Britton, L., et al. 1975. An Introduction to the Processes, Problems, and Management of Urban Lakes. Geological Survey circular 601-K. U.S. Geological Survey, National Center, Reston, VA 22092.

12

Nitrate in Groundwater

Nitrogen is a fairly inert gas. About 78% of the air we breathe is nitrogen. A variety of algae, bacteria, and fungi can "fix" inert nitrogen gas into chemicals that they can use (or that organisms that associate with them or feed on them can use) to manufacture their proteins. When an organism dies, the nitrogen in its decomposing protein is released to the environment in various chemical forms, including nitrogen gas. The sum of processes involving fixation and release of nitrogen gas is called the "nitrogen cycle." Detailed descriptions of cycle processes can be found elsewhere.[1]

Nitrate, one of the chemical compounds in the nitrogen cycle, is of interest to us. The U.S. Public Health Service and the World Health Organization drinking water standard for nitrate is 10 mg/L as nitrate-nitrogen (45 mg/L as nitrate).*

---

*The 10 mg/L standard was established to prevent methemoglobinemia ("blue baby" disease) in infants. Ingested nitrate combines with hemoglobin (the red pigment in blood that transports oxygen and carbon dioxide to and from the cells in our body). When hemoglobin combines with nitrate, it becomes methemoglobin, and cannot do its job. Babies less than 3 months old (and some adult Eskimos) lack a reducing enzyme that can knock the nitrate out of the methemoglobin; when they ingest too much nitrate they turn bluish, and may die.

In the last few years, the amount of nitrate used as a preservative (in hams, hot dogs, corned beef, and other processed meats) has been lowered due to another type of concern: cancer. The probability is low that any single individual will contract cancer due to ingested nitrates. But when a low individual risk is shared within a large population, some members of this population may contract cancer.

145

This standard may be exceeded in groundwaters affected by septic system discharges.

Sewage effluent discharged into septic tanks usually has only a trace of nitrates. About 1/4 to 1/3 of the nitrogen in sewage is part of fecal protein (and fecal amino acids and amino sugars); the rest is part of urea (urine's main nitrogen compound). In the septic tank or in the soil, a bacterial enzyme splits urea into ammonia and carbon dioxide. Fecal proteins, when decomposed, may also release ammonia. Ammonia combines with water to form ammonium ions. Aerobic bacteria oxidize the ammonium ion to nitrate, usually within 2 to 5 feet of unsaturated (aerobic) soil,[2] below the leachfield. This oxidation is called nitrification. If bacteria have carbonaceous food available* and their environment is anaerobic (or if oxygen doesn't reach them fast enough when they are "eating" the carbonaceous food), they use the nitrate's oxygen atoms instead and reduce nitrate to nitrous oxide and to molecular nitrogen. Then, these two gases escape into the atmosphere. This is nitrogen loss by denitrification.

In the next sections we will explore somewhat controversial topics: how much (if any) nitrate is generated in a leachfield, how much of it is denitrified, and how nitrate pollution is controlled.

12.1 A SEPTIC SYSTEM'S NITROGEN INFLOW AND OUTFLOW

The per capita ingestion and excretion of N (nitrogen) are fairly constant, with relatively minor and predictable variations due to diet, weight, and age. The per capita intake of N/day can form the basis for estimating nitrogen inflows and outflows. According to Carla Bouchard, R.D., a public health

---

*Sugars, organic acids, alcohols, aldehydes, molasses, or any other biodegradable matter that increase the carbon-nitrogen ratio to more than 1.3:1.

nutritionist (personal communication), the actual consumption of protein is between 70 and 140 g/capita/day (nationwide average), twice as much as necessary.

Nitrogen constitutes 16% of the weight of the average protein. Hence, 16% of 70 and 140 gives us, respectively, 11 and 22 g N/capita/day. For comparison purposes, Laak[3] has referred to an EPA study (Process Design Manual for Nitrogen Control, 1975) and quoted a figure of 11 g N/capita/day; the EPA Manual figures (Table 4–3 in Reference 4) are 6 to 17 g N/capita/day.

A negligible proportion of ingested N is incorporated into growing bodies.* Some of the nitrogen excreted stays in (and is pumped out of) the septic tank, but this small amount can be neglected as well.**

Most literature evaluates nitrogen discharged from septic tanks by means of concentration units, mg N/L.[4-9] This procedure superimposes flow variability onto nitrogen excretion variability, so it is not surprising that one researcher might measure a mean of 35 mg N/L[4] and another a mean of 97 mg N/L.[9] Bauman[7] has referred to one of his previous studies in which he averaged the means of 20 other studies and obtained a concentration of 62 ± 21 mg/L nitrogen in septic tank effluent; he did not state whether the other 20 studies were an unbiased sample of the United States as a whole.

Now let's see how 11 to 22 g/capita/day compares to the figures in the literature. The average per capita discharge of sewage is 64 gallons/day, according to Ingham,[8] and 55 gallons/day per (some portions of) the EPA Design Manual;[4]

---

*A youngster who gains 5 kilograms in one year retains as much nitrogen as is contained in the food he eats in 5 days.

**Septage (liquid pumped from a septic tank) contains about 588 mg/L of total N (average of 4 studies with means of 510, 572, 650, and 820 mg/L, as reported in Table 9–2 of the EPA Design Manual[4]). For an average family of 3 people served by a septic tank of 3,800 liters capacity pumped every 3 years, the figures are: Total grams of N discharged in 3 years = $3 \times 365 \times 3 \times (11 + 22)/2 = 56,000$. And total grams of N lost in pumped septage = $3,800 \times 588/1,000 = 2,200$, only 4% of the total discharged.

let's say 60 gallons/day, or 230 liters/day. Dividing 11 and 22 g by 230 L, we obtain nitrogen concentrations of 48 and 96 mg/L, respectively—say roughly 72 mg/L, which is not too far from Bauman's figure. (And it is even closer if we assume that 15% of the N is lost from the septic tank, as in Andreoli et al.'s experiment.[2] But this loss has not been verified with replications.)

Now, let us try to figure out the magnitude of nitrogen loss when septic effluent percolates through the soil. Winneberger assumed that denitrification was a major process, and stated that an 85% loss of N was a conservative figure.[9] But Sikora, et al. asserted that "passage of septic tank effluent through a seepage field [leachfield] usually results in a nitrified effluent with insufficient BOD [carbonaceous food] levels to support denitrification."[10] And Perkins observed that ". . . data relating chemical contamination [of groundwater], especially nitrate contamination, to septic tank density are becoming numerous" and mentioned various instances and nine pertinent publications.[6]

Eastburn and Ritter reviewed fairly current literature that included reports of up to 48% to 86% denitrification (in some mound systems), and concluded that "denitrification rates under conventional septic systems may vary from 0 to 35%"* and that "absolute values of denitrification for conventional septic systems cannot be predicted based on the available literature."[11]

From the above, we may surmise that, for gross planning or evaluation purposes, we should neglect denitrification

---

*This conclusion is in harmony with views in literature not cited by Eastburn and Ritter. Regarding wastewater irrigation, Broadbent and Aref[12] found up to 32% denitrification loss when sewage percolated through aerobic soil columns. And Broadbent and Reisenauer[13] stated that coarse-textured, well-drained soils with low organic matter content have negligible denitrification potential; sandy loam and loam soils have medium denitrification potential (10 to 20% N loss); and finer textured soils, high potential[2] (20 to 40% N loss).

losses and use two parameters: $(11 + 22)/2 = 17$ g N/capita/day, and Bauman's 62 mg N/L of discharged effluent.

12.2 WHAT IS BEING DONE

There is an incredible variety of technologies to denitrify sewage. Practically all are based on promoting nitrification of effluents and then supplying carbonaceous food to the bacteria in the effluent in an anoxic medium. The carbonaceous food may be sewage BOD constituents, or methanol, or whatever. The medium may be a special chamber or the soil itself. Almost total nitrogen removal is possible; but such removal may not be practical in every case. Prakasam and Krup[14] have given an extensive review of technological approaches. (Among their 122 references, more than 2 dozen refer to denitrification processes patented in Japan.) The EPA has published a detailed description of basic nitrogen control technologies.[15]

In addition to technological approaches to control nitrate pollution, there are land use approaches. There are three main land use approaches for preventing or minimizing degradation of groundwater. One is to prevent development in watershed or groundwater supply areas. Another one is to install sewers and discharge treated sewage pollutants somewhere else where their effect is not too deleterious. And the last one is to control the density of septic systems, i.e., to restrict use of septic systems to lots of a given large size.

In southern California, as a means to minimize degradation of water in the Santa Ana river basin by TDS or "salts," a minimum lot size of one-half acre is required in large residential developments. In northern California, some counties require minimum lot sizes of five acres or larger for various "environmental" reasons. Perkins has summarized pretty well the state-of-the-art knowledge re controlling density of septic systems and nitrate pollution.[6] However, I think he is much too optimistic about the general applicability of the mathematical models he discussed.

Most of the models Perkins described[6] are too simplistic to account for a wide variety of field conditions. If and when they "work," the models could show that an area as a whole does not suffer from nitrate pollution, yet many wells within this area could be affected. On the other hand, another model mentioned by Perkins,[6] the "U.S.G.S. Method-of-Characteristics Model for Solute Transport of Groundwater" (Konikow-Bredehoeft model), is complex enough to take into account many pertinent variables. But complex models present nearly insoluble problems. Canter and Knox tried out the Konikow-Bredehoeft model.[5] Excerpts from their evaluation follow:

> The results . . . must be classified as disappointing and frustrating . . . the model was unable to be calibrated for groundwater flow . . . a parameter determined from a particular test in a particular spot may be accurate but may not reflect the gross properties of the aquifer. The only conclusion to be drawn concerning the applicability of sophisticated ground water models to the problem of septic tank systems is that the utility of the models may be outweighed by their significant data requirements.

I am in complete agreement.

Although we do not have practical and reliable tools to determine specific minimum lot sizes, we can make rough estimates on the basis of data in the Perkins[6] and Bauman and Schafer[7] articles. Perkins stated,[6] "The range of [minimum] lot sizes . . . appears to be from 1/2 to 1 acre based on reported data, and from 3/4 to 1 acre, based on theory." But Bauman and Schafer[7] calculated that the nitrate standard would be exceeded if the lots were less than 1 to 2 acres and the groundwater moved less than 31 m/year (hydraulic conductivity of 3,154 m/year and 1% gradient [or hydraulic conductivity of 315 m/year, and 10% gradient], 40-acre development, effluent 62 mg N/L, 186,500 L of effluent/family/year,

10 m mixing zone on top of the aquifer). So, as a "first approximation," we may say that, for an isolated 40-acre development with onsite wells, nitrate pollution could be a cause for concern if the lots are less than about 1.5 acres in size and/or groundwater moves slower than 31 m/year (100 feet/year).

12.3 DEALING WITH A SPECIFIC PROBLEM OF NITRATE POLLUTION

Caution: To safely deal with real nitrate pollution problems, one should know a bit of hydraulics and geology and should read in depth (and with a critical eye) the references given in this chapter. What follows is a mere sketch, not a methodology or prescription.

The first thing one should do is to define what is meant by nitrate pollution in the specific setting: What will be the "source" (i.e., amount, type, location of septic systems), and what will be the "sink" (location and depth of well or wells; or, will the nitrate plume go to marshlands, agricultural areas, the ocean, etc.).

The second thing is to learn as much as possible about the specific setting. Cannery wastes are high in BOD (carbonaceous food for bacteria) and could be used to denitrify collected septic tank effluent (a sanitation district would have to be formed). If the nitrate plume will pass through a marshland, the nitrates might be consumed or denitrified. If the plume goes to a groundwater basin serving an agricultural area, it may pay to know whether the impact will be worthy of note. While farmers apply about 100, 300, or more pounds of N per acre per year, and about half of this nitrogen is leached down to the groundwater as nitrate, the septic system of a household with 3 people discharges only about 40 pounds of nitrate-N per year. If a subdivision's lots are large enough to allow a "gentleman farmer's lifestyle," more nitrates may be generated by lawn, tree, or crop fertilization than by septic systems.

The third thing is to estimate the flows of water and of nitrates, and to approximate a worst-case situation. One cannot predict exactly what the level of pollution will be, and exactly where it will occur. But one can make conservative assumptions, like zero denitrification in the soil, and assume the worst case. And if the worst case is not too bad, there is no need to worry. An illustration is given below.

Years ago, I had to determine whether a high-density development served by seepage pits would result in nitrate pollution of a well nearby, as follows. (The actual figures may be different.) I assumed a generation of 6 kg of N (as nitrates) per year per each 1 of the 50 residents. The water table was at 200 feet and the well's perforated casing extended from 300 to 400 feet. The groundwater was practically nitrate-free and moved at a probable rate of at least 1 foot per day. I did not know (no one knows) the depth of the mixing layer where percolated sewage supposedly mixes with groundwater.* But by assuming that the well extracts a volume of water from 200 feet down to 400 feet, and over the width of the development (perpendicular to groundwater flow) of 200 feet, the thickness and concentration of the mixing layer is irrelevant: The well mixes (whatever amount of) nitrate (is in the upper polluted water layer) with all the clean water pumped to a depth of 400 feet. So, we have a "cube" (parallellepiped) of water 200 feet wide, 200 feet deep, and at least 365 feet long, which receives about 300 kg of nitrate-nitrogen per year. The total weight of water in this cube is 400 million kg. Hence, the nitrate concentration is

---

*By the way, it may be risky to assume given thicknesses of (or concentrations in) the mixing layer. In semi-arid climates, nitrates may stay above the groundwater during some years and then leach down in mass during other years, depending on the rainfall regime and soil conditions. In shallow leachfields, nitrates may move down with rainfall and then up by capillarity and evapotranspiration. Soils are capable of performing all kinds of tricks just to confound experts.

300/400=0.75 parts per million or 0.75 mg N/L. The well wasn't even in the presumed path of the pollution plume, and the plume probably had a tendency to stay over the top of the groundwater because of a bit of stratification in the aquifer. So, it was a safe bet that the 10 mg N/L standard would not be exceeded. (And a bet it was. Among other strange things, in California we have geologic faults all over the place and plumes may not move the way they are supposed to.)

And finally, the last thing to do is to compare and choose among the lowest cost strategies that are viable. There are hundreds of possible ways to match technical and land use approaches. It is beyond the scope of this book to delve into these.

REFERENCES

1. Delwiche, C. (Ed.) 1981. *Denitrification, Nitrification, and Atmospheric Nitrous Oxide.* J. Wiley & Sons, New York, NY.
2. Andreoli, A., et al. 1979. Nitrogen removal in a subsurface disposal system. *J. Water Poll. Control Fed.* 51:851–855.
3. Laak, R. 1982. A passive denitrification system for on-site systems. In ASAE Pub 1–82, pp. 108–115. ASAE, St. Joseph, MI.
4. U.S. EPA. 1980. Design Manual. EPA 625/1–80–012.
5. Canter, L.W. and Knox, R.L. 1985. *Septic Tank System Effects on Ground Water Quality.* Lewis Publishers, Chelsea, MI.
6. Perkins, R. 1985. Septic tanks, lot size and pollution of water table aquifers. *J. Environ. Health* 56:298–303.
7. Bauman, B.J., and Schafer, W. M. 1985. Estimating groundwater quality impacts from on-site sewage treatment systems. In ASAE Pub. 07–85, pp. 285–295. ASAE, St. Joseph, MI.

8. Ingham, A. 1980. Residential Greywater Management in California. (It quotes data from "Water Conservation in California," Bulletin 198, May 1976.) State Water Quality Control Board, Sacramento, CA.

9. Winneberger, J.T. (Undated. The last reference is dated 1971; probably published on or about 1972.) Hancor's "On-Site Waste Management," vol II. Hancor, Findlay, OH.

10. Sikora, L., et al. 1976. Septic nitrogen and phosphorus removal test system. *Ground Water* 14:309–314.

11. Eastburn, R.P., and Ritter, W. F. 1985. Denitrification in onsite wastewater treatment systems—a review. In ASAE Pub. 07–85, pp. 305–313. (See reference 7 above.)

12. Broadbent, F.E., and Aref, K.E. 1976. Nitrification and Denitrification of Municipal Wastewater Effluents Disposed to Land. Publication No. 58. California State Water Resources Control Board, Sacramento, CA.

13. Broadbent, F.E., and Reisenauer, H.M. 1984. Fate of wastewater constituents in soil and groundwater: nitrogen and phosphorus. In "Irrigation with Reclaimed Municipal Wastewater." Report No. 84–1wr. California State Water Resources Control Board, Sacramento, CA.

14. Prakasam, T.B.S., and Krup, M. 1982. Denitrification. *J. Water Poll. Control Fed.* 54: 623–631.

15. Brown and Caldwell. Process Design Manual for Nitrogen Control. 1975. EPA 625/1-75-007. NTIS Report no. BA PB 259/49/LS.

13

Mounding

If sewage flows downward through a porous medium and is stopped or slowed down by a barrier, it accumulates above the barrier and forms a mound of saturated soil. As the height of the mound increases, the hydrostatic pressure builds up and pushes the sewage laterally.

When a mound forms below a leachfield (leachline, seepage pit, seepage bed, soil mound, etc.), the separation between the bottom of the leachfield and groundwater (or saturated soil) may decrease to less than minimum code requirements. The mound may grow high enough to flood the leachfield. And, of course, saturated soil (within the mound) facilitates the movement of pathogens to the groundwater table.

Mounds may form over two kinds of "barriers." One kind is clay, caliche, or other low-permeability stratum; the other is a groundwater table's surface.

13.1 MOUNDING OVER A LOW-PERMEABILITY STRATUM

Early in 1985, I reviewed a perk report in which seepage pits 20 feet deep were recommended for a high-density development. At the site, a loam surface soil graded to clay loam with depth. There was a clay stratum (and traces of perched groundwater) at a depth of 30 feet. (The code requires a minimum separation of 10 feet between bottom of

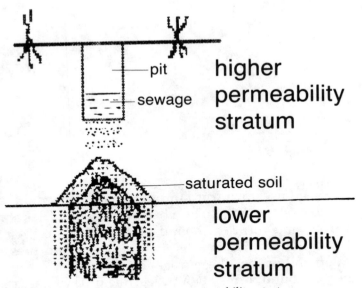

pit

sewage

higher
permeability
stratum

saturated soil

lower
permeability
stratum

Figure 13.1   Mounding over a lower-permeability stratum.

pit and impermeable layer or groundwater.) I doubted that
the recommendation was appropriate because of the effect of
mounding. Therefore, I derived simple, practical formulas to
estimate mounding, and showed that the heights of the
mounds above the clay stratum could exceed 20 or even 30
feet. The top of the mounds could have been near the top of
the pits, and sewage would not have been absorbed. So the
proposed density of development was halved, and leachlines were used instead of pits.

Derivations of formulas to estimate mounding under ideal
conditions follow. Ideal conditions are that the soil strata are
isotropic (and homogeneous) and horizontal, and the saturated flows obey Darcy's formula. Sewage percolates down,
encounters a lower-permeability stratum, and mounds over
it. Figure 13.1 depicts mounding over a lower-permeability
soil.

13.1.1 Mounding Under a "Line Source" (Leachline)

Let us visualize a series of time-lapse photographs of viscous syrup trickling onto a plate. After impact, the first drop forms a little cone that flattens out and spreads to the extent permitted by the viscosity of the syrup, and then turns into a flat cylinder. The second drop forms a little cone over the top of this cylinder; as the weight of the cone pushes down, the diameter of the cylinder below increases, and the cone itself subsides and spreads into a flat cylinder, on top of the previous one—and so on. The faster we pour, the higher the cone. The higher the cone, the more pressure is exerted by its weight downward and the bigger becomes the base of the cone.

Now let us visualize something similar. However, this time the syrup is poured between the middle of two vertical, parallel glass panes, half an inch apart. We see a solid triangle of syrup form (between the glass panes), as if it were a cross-section through the middle of a cone of syrup.

Now, let us imagine that we place a fine wire mesh under the glass panes, horizontally, and we allow the syrup to ooze slowly through it. (This is represented in Figure 13.2.) As we pour syrup between the glass and over the mesh, we see that the solid triangle increases its height and base until the amount of syrup going into it is equal to the amount of syrup oozing out through the mesh below. This is called "steady state."

Let us modify this last experimental fantasy: let us imagine that water percolates down a soil between vertical, parallel glass panes until it hits a low-permeability layer. It also forms a triangle between the glass panes. The force that pushes the water laterally is the weight of water above it; the forces that counteract this push are viscosity and friction in the soil channels. In Figure 13.2, the triangle between the two glass panes represents a vertical cross-section through a mound below a "line source" (leachline).

Now, the derivation follows. Figure 13.3 shows a steady-

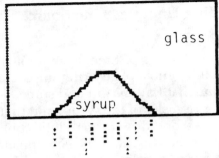

Figure 13.2 Triangle of viscous syrup over a mesh.

state triangle within a soil. The "line source" lies just above the triangle and runs parallel to the Z axis. The "barrier" is the lower-permeability stratum below the X-Z plane. The daily flow is Q. Half of it goes to the left of the Z axis, and half to the right. The flow 1/2 q through a small element of the triangle with height dh follows the general Darcy formula shown in Equation 1. The gradient is proportional to the difference between hydraulic pressures at two points, the "left" and the "right" ends of this element (respectively h and 0), divided by X. (Note that at the top of the triangle, hydraulic pressure = 0; at the bottom, it is proportional to h.)

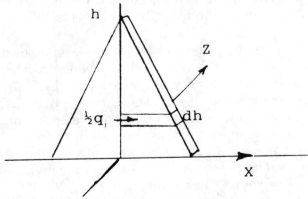

Figure 13.3 Steady-state triangle within a soil.

The thickness of the triangle is 1 foot—i.e., $Z = 1$ foot. Steady-state conditions exist.

Per Darcy's formula,

$$\text{ft}^3/\text{day}/\text{linear foot} = k(\text{ft}/\text{day}) \times H/X \times A \ (\text{ft}^2) \tag{1}$$

$$\text{Let } X = ah \ (\text{"a" is an unknown proportionality constant}) \tag{2}$$

$$\tfrac{1}{2}q_i = k(h_i/ah_i) \ dh = (k/a) \ dh \tag{3}$$

$$\tfrac{1}{2}Q = \tfrac{1}{2} \Sigma q_i = (k/a)\int_0^h dh = kh/a \tag{4}$$

From Formula 2, $a = X/h$, and substituting this value in Formula 4,

$$\tfrac{1}{2}Q = \tfrac{1}{2}kh^2/X \tag{5}$$

Per Darcy's formula, the saturated flow downward out of the half-triangle is $1/2 \ Q/A = k' \times 1$. A is the area at the base of the half-triangle, equal to X times the width; the width is 1 foot $(Z = 1)$. k' is the hydraulic conductivity of the lower-permeability stratum. And 1 is the value of the gradient downward: we assume that the flow has occurred over a long period of time, so that the column of saturated soil below the triangle is very large. Hence,

$$X = \tfrac{1}{2}Q/k' \tag{6}$$

And substituting this value of X in Formula 5,

$$\tfrac{1}{2}Q = (\tfrac{1}{2}kh^2)/(\tfrac{1}{2}Q/k') \tag{7}$$

And rearranging terms and canceling out constants, we obtain the value of h:

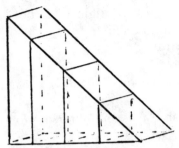

Figure 13.4 Vertical surfaces through which sewage flows downward.

$$h = \sqrt{Q^2/2kk'} \tag{8}$$

There is another way of arriving at the same result. The solid triangle in Figure 13.4 shows vertical surfaces through which the sewage flows ("to the right") before it moves downward. The extent of flow obeys Darcy's formula; it is proportional to the area of the vertical surface and to the gradient, here expressed in differential form, dh/dx.

Each of these vertical surfaces measures "h feet" in height and 1 foot in width. Temporarily assuming that the bottom is impermeable, so that the Q through each vertical surface is about the same, we have:

$$\tfrac{1}{2}q_i/(h_i \times 1) = k \; dh/dX \tag{9}$$

$$\tfrac{1}{2}Q = k \times h \times dh/dX \tag{10}$$

$$\tfrac{1}{2}Q \; dX = kh \; dh \tag{11}$$

Integrating Formula 11 from X = 0 to X = X and from h = 0 to h = h, we obtain Formula 12.

$$QX/k = h^2 \tag{12}$$

and as we saw in Formula 6, X must be equal to ½ Q/k'. Substituting this value of X in Formula 12 and rearranging, we obtain

$$h = \sqrt{Q^2/2kk'} \qquad (13)$$

which is identical to Formula 8.

Let us try the formulas out. Let us assume that a leachline discharges 1 cubic foot of sewage per day per lineal foot, and that k = 5 ft/day and k' = 0.01 ft/day. Then,

$$h = \sqrt{1^2/(2 \times 5 \times 0.01)} = 3.2 \text{ ft}$$
$$X = \tfrac{1}{2}Q/k' = 0.5/0.01 = 50 \text{ ft}$$

Generally, a mound with a base of 2X = 100 feet, and as long as the leachline plus a few more feet at each end, may never reach full theoretical growth, because it grows slowly and loses water to unsaturated flow, to evapotranspiration (see Section 5.5.1), and to drainage (over and under a sloping low-permeability barrier).

Assuming that the soil has 50% voids, the time for 1 cubic foot per day (per lineal foot of leachline) to fill the voids and saturate 30 feet of soil below the solid triangle in the previous problem (so that the downflow gradient becomes close to unity) is 0.5 × 100 × 30 = 1,500 days; and to fill the solid triangle, 0.5 × 0.5 × 100 × 3 = 75 days.

13.1.2 Mounding Under a "Point Source" (Seepage Pit)

Let us imagine cylindrical surfaces within a cone, as shown in Figure 13.5. Let us use a derivation similar to that leading to Formula 13 above. The difference is that the vertical surfaces are cylindrical, and the area through which Q flows down and out of the cone is circular.

Then, per Darcy's formula,

$$Q = Sk(dh/dr) \tag{14}$$

where dh/dr is the gradient in differential form and S is the area of the cylindrical surface. So,

$$Q = 2\pi rhk(dh/dr)$$

Rearranging terms, we obtain

$$Q \, (dr/r) = 2 \, \pi khdh \tag{15}$$

and integrating this from $r = r$ to $r = R$, where R is the maximum radius at the base of the cone; and from $h = 0$ to $h = H$, where H is the maximum height of the mound, we obtain

$$Q \ln(R/r) = \pi kH^2 \tag{16}$$

The value of R corresponds to the maximum basal radius of the cone, and the value of r to the radius at the top of the mound (roughly equal to the radius of the pit). The cone grows until the flow downward out of the base of the cone is equal to the flow coming into the cone from above. After a

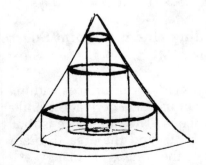

Figure 13.5 Model for mounding under a point source.

long time, the gradient of the flow downward through the low-permeability clayey stratum below the base of the cone is close to unity, because the ratio $(h + X)/X$ approaches unity (see Chapters 5 and 6). So the flow through the base of this cone is $Q/A = k' \times 1$, where A is equal to the area at the base of the cone, k' is the hydraulic conductivity of the clayey stratum, and 1 is the value of the gradient. And since $A = \pi R^2$, we have that at steady state

$$Q = \pi R^2 k' \qquad (17)$$

Hence,

$$R = \sqrt{Q/\pi k'} \qquad (18)$$

and plugging this into Formula 16, and neglecting a small amount of flow downward through the area πr^2, we have that

$$Q \ln[\sqrt{(Q/\pi k')}/r] = \pi k H^2 \qquad (19)$$

And solving for H,

$$H = \sqrt{(Q/\pi k)\ 1.15 \log(Q/\pi k' r^2)} \qquad (20)$$

Let us try out this formula.

A seepage pit is installed in a very permeable sandy soil, but 10 feet below the bottom of the pit there is a thick stratum of less permeable loam soil. The pit receives 300 gallons of sewage per day (40 cubic feet). Let us assume that $k = 5.2$ and $k' = 0.1$ ft/day. If $r = 2.5$, what is the maximum possible value of H? Of R?

$$H = \sqrt{[40/(3.1 \times 5.2)] \times 1.15 \log[40/(3.1 \times 0.1 \times 6.25)]}$$
$$= 1.9 \text{ feet}$$

$$R = \sqrt{40/(3.1 \times 0.1)} = 11 \text{ feet}$$

Again, same as above, but k = 0.1 and k' = 0.01 ft/day.

$$H = \sqrt{[40/(3.1 \times 0.14)]} \times 1.15 \log[40/(3.1 \times 0.01 \times 6.25)]$$
$$= 18 \text{ feet}$$

$$R = \sqrt{40/(3.1 \times 0.01)} = 36 \text{ feet}$$

And now with k = 1 and k' = 0.001,

$$H = \sqrt{[40/(3.1 \times 1)]} \times 1.15 \log[40/(3.1 \times 0.001 \times 6.25)]$$
$$= 7 \text{ feet}$$

$$R = \sqrt{40/(3.1 \times 0.001)} = 113 \text{ feet}$$

13.1.3 Estimates of Maximum Mounding Heights

I used the previous formulas to estimate the maximum possible height of mounding seen in Table 13.1 by assuming that:

a. A leachline with 7.5 square feet of absorption area per lineal foot receives Q's (or sewage flows) per loading rates recommended in Table 7–2 of the EPA Design Manual.[1]
b. The relationship between k and mpi is as determined by Winneberger in decomposed-granite soil.[2] Presumably, this decomposed granite was fairly isotropic and approximated an ideal soil. The relationship between k and mpi changes a bit in other types of soil, according to Fritton et al.[3] This is to be expected, as most soils are not isotropic.

In a horizontally stratified soil, k in the horizontal direction may be 100 times larger than in the vertical direction. The horizontal permeability constant corresponds to k, and the vertical one to k'.

It should be emphasized that the permeability constants of clayey soils are very variable and unreliable. Although two volume units or "lumps" of soil may be identical in practically all respects (including density and total pore volume), if one lump has its pore volume concentrated into a just a few macropores, its k will be much higher than the other lump's k. Also, even nonstratified soils might have very different k's for flows in the vertical and in the horizontal direction, according to the direction and continuity of the macropores. Therefore, we must remember that the formulas were derived for ideal, isotropic strata.

Table 13.1   Maximum Height of Mounding[a]

soil type	mpi	k	leachlines Q(L)	leachlines h	leachlines X	seepage pits Q(Pit) 300 h	seepage pits Q(Pit) 800 h	Q(UPC)
s	3	5.2	1	7	500	3	5	5 (60)
a l	30	0.14	0.5	21	250	19	32	5 (6)
n o c	45	0.08	0.5	28	250	25	43	4
d a l	120	0.02	0.2	112	100	50	85	1.5
m a	163							1.1
y	220							0.83
	600	0.0014						0.3

[a]The soil types are very approximate. k is as given, in ft/day; k'=0.001 ft/day. h = feet [maximum height of mound above the low pemeability layer]. Q(L) [for leachlines] is in ft^3 per ft of leachline length. X is in ft. Q(Pit) represents two pit loading rates, 300 and 800 gal/day [66 and 110 ft^3/day, respectively]; the corresponding values of R are 113 and 187 feet. Q(UPC) is in gal/ft^2/day; this column translates mpi to the requirements of Table I-4 of the Uniform Plumbing Code[4] by means of the relation Q [gallons of sewage/ft^2/day] = 180/mpi; the UPC's maximum allowable Q is 5 gallons of sewage/ft^2/day. (For comparison purposes: Per California Administrative Code Title 23 [Sections 2533 and 2542], clays used as liners to seal landfills must have a k lower than 0.003 feet/day.)

13.1.4 Discussion of Mounding Over a Low-Permeability Stratum

From Table 13.1 and the formulas derived, we can conclude that:

a. The maximum height of mounding increases as k decreases.
b. The base of the mound increases as k' decreases.
c. Many soils have alternating strata of low and high permeability. When a seepage pit is installed in such soils, the top of the mound may be within the pit itself.
d. When many pits are installed over a low-permeability stratum, one should ensure that the R's do not overlap too much. The UPC's minimum separation of 12 feet between pits may be inadequate unless the soil is deep and very permeable, or the loading rate is low.
e. Usually less than 500 gallons of water are used to perform a falling head test for a seepage pit. (The test hole is 6 to 10 inches in diameter.) And during this test the h is "full height," but the parameters that decrease the gradient (i.e., r and depth of wet soil below the bottom of the test hole) are only a fraction of their full magnitude at steady state. (For instance, at near steady state, the volume of water in the wetted soil when R = 113 feet is well in excess of 500 million gallons.) So, the gradients during testing are much larger than during true equilibrium or steady state conditions, and the measured Q (absorption rates) should be much larger than at equilibrium. But very often the compaction of test hole sidewalls is substantial and the rate of absorption is reduced drastically. Hence, all in all, the raw results of seepage-pit falling head tests are a poor indicator of required size or expected longevity.

In practice, most soils used for sewage disposal have (saturated) k values of about 0.1 to 3 feet/day, and k' values of 0.01 feet/day or higher. Mounding is not a common occurrence under leachlines because of the large horizontal spread of the sewage and the influence of evapotranspiration, drainage, and unsaturated flow. Mounding may be more of a problem in pits installed in some stratified soils with low k and k' values, and especially so if pits are concentrated within a small area; even then, it may take more than 4 to 8 years for the top of the mound to rise to the top of a pit's inlet. By this time, the actual absorption of sewage might have decreased because of formation of a clogging layer, and mounding might subside if the flow goes to other faraway pits in series. Where pits 100 feet deep are used, it may take more than 100 years before steady state is reached and mounds reach maximum height. If one must have a pit installed in low-permeability soil, it is much better to have one 100-footer than two 50-footers.

13.2 MOUNDING OVER A GROUNDWATER TABLE

Formulas to calculate mounding over a groundwater table were derived by Hantush in 1967.[5] They were simplified by Finnemore and Hantzsche in 1983,[6] and, according to Canter and Knox,[7] by Kincannon in 1981.

Canter and Knox[7] presented Kincannon's very simple step-by-step procedure to calculate mounding, and provided an extensive table of necessary "W" coefficients. It is not proper to repeat in this book what was well presented elsewhere.[7] However, some important details were left out of Canter and Knox's book, and are noted below:

a. The procedure may not yield accurate results if the height of the mound exceeds 50% of the original depth of saturated soil.[5] In other words, if the thickness of an aquifer is 100 feet, and the expected mound over the top of this aquifer will be more than 50 feet high, the calculated height may be in error by more than 6%.[5] But Finnemore and Hantzsche[6] quote other investigators who reported that the error was less than 2% when mounding raised groundwater level more than 3 times the initial saturation depth, under a line source.

b. Transmissivity (in gallons/day/foot) is calculated by multiplying hydraulic conductivity k (in gallons/day/square foot), by the thickness of the aquifer (in feet).

c. "Average" specific yields of various aquifer materials can be found in Figures 3.3 and 3.4, page 3–9, of the EPA Process Design Manual for Land Treatment of Municipal Wastewater.[8] These rough averages can be used for planning or comparison purposes; the calculated heights of mounding are not too sensitive to errors in specific yields.[6]

d. Step 5 (of the procedure) calls for adding together the four "W" values obtained in Steps 3 and 4.

e. The "W" table in their book (Reference 7, Appendix D, pp. 263–261) was not specifically labeled as such.

f. It is cumbersome to calculate mounding heights for times longer than a year or so. Sometimes the coefficients in the "W" table are not precise enough. But Finnemore and Hantzsche have solved this problem (see their Formula 4).[6]

Finnemore and Hantzsche[6] made practical observations about mounding above an aquifer:

a. An average individual home discharging 250 gallons/day to leachlines, seepage beds, or soil mounds within an area of 600 square feet may raise groundwater not more than about 13 inches. (But, looking at Figure 4 of their paper, one can estimate that over 10 years the mound height can exceed 12 feet, if the aquifer below the mound is less than 10 feet thick and the k is less than 1 ft/day.)

b. When trying to maximize separation between leachline bottom and groundwater, there is little advantage to alternating two leachfields instead of continuously discharging half the flow to each leachfield.

13.3 PROBLEMS

Estimate the maximum possible height of mounding under ideal conditions. Use k = 0.1, k' = 0.01 feet/day.

a. One pit receives a Q of 160 cubic feet per day.
b. Three pits are 12 feet apart from each other (in a triangular pattern). Each receives 40 cubic feet per day.

REFERENCES

1. U.S. EPA. 1980. Design Manual. EPA-625/1-80-012.
2. Winneberger, J.T. 1974. Correlation of three techniques for determining soil permeability. *J. Environ. Health* 37: 108-118.
3. Fritton, D.D., et al. 1982. A site evaluation model for effluent disposal. In ASAE Pub. 1-82, pp. 32-41, ASAE, St. Joseph, MI.
4. IAPMO. 1985. Uniform Plumbing Code. IAPMO, Los Angeles, CA.
5. Hantush, M. S. 1967. Growth and decay of groundwater-mounds in response to uniform percolation. *Water Res. Res.* 3:227-234.
6. Finnemore, E.J., and Hantzsche, N.N. 1983. *J. Irrig. Drainage Eng.* 109:199-210.
7. Canter, L.W., and Knox, R.C. 1985. *Septic Tank System Effects on Ground Water Quality.* Lewis Publishers, Chelsea, MI.
8. U.S. EPA. 1981. Process Design Manual for Land Treatment of Municipal Wastewater. EPA-625/1-81-013.

14

Septic Systems, Sewers,
and Land Use

Because of alleged "failure" problems, septic systems have a somewhat tarnished reputation. Of course, it is not septic systems that fail—it is people who fail to design, locate, install, and use them correctly.

What is not so well known is that sewers also "fail." Rae Tyson wrote* about the contents of a 1986 draft by an EPA advisory group. An abstract from the article follows:

> Half the nation's major sewer plants violate the Clean Water Act by discharging pollutants into lakes, rivers, streams, and oceans. About 3,360 of some 12,000 smaller plants (capacity less than 5 million gallons per day) also violate effluent standards on a regular basis. 46% of all major municipal permittees are in serious violation of the Act.

Septic systems are not "better" than sewers. Sewers are not "better" than septic systems. Each has its advantages and disadvantages.

14.1 AVAILABILITY

Where sewers are not available, septic systems are usually the only practical alternative. Where septic systems cannot be installed due to site constraints (high groundwater table,

---

*Gannett News (*The Sun*, San Bernardino County), January 19, 1986.

solid-rock ground, groundwater quality protection, etc.), sewers are the preferred alternative.

14.2 CONVENIENCE

Sewers are extremely convenient: "just flush and forget." A sewered house may discharge a sewage volume 10 times higher than normal with no problems. The design of sewers and treatment plants is fairly standard.

The use of septic systems requires some knowledge about the limitations of the system: what and how much can be flushed, and how often to pump the septic tank. Discharges much larger than normal may cause premature failure. The design of septic systems ought to be sensitive to site conditions, but this is often not the case.

14.3 COST

As a very rough approximation, the present-worth cost of sewering is 1 to 9 times that of septic systems. Sewering becomes cost-competitive if lots are smaller than 0.5 acres. Precise figures cannot be given because too many local variables affect the cost figures. Details about costs can be found elsewhere.[1,2]

14.4 HEALTH AND POLLUTION HAZARDS

Sewers carry away large amounts of pollutants, parasites, and pathogens for treatment and discharge at remote offsite locations where harmful effects may be minimized. Septic systems treat effluents onsite; nitrates, rarely phosphates, and extraneous synthetic chemicals may contaminate local groundwater. Both sewers and septic systems can overflow and create health hazards. The magnitude and location of these hazards are obviously different.

Rats, which are vectors of various diseases, may live in, propagate in, and spread through an urban area's sewer network.

14.5 LAND USE AND ENVIRONMENTAL IMPACT

Sewers are the result and the cause of high-density development. Availability of sewers encourages development, especially high-density. The resulting growth may entail air pollution, congestion of traffic arteries (especially in mountain communities with narrow winding roads), and pollution of surface waters due to stormwater runoff. Urban stormwater runoff is highly polluted.[3] It contains lead (from car exhaust), parasites and pathogens from pet feces, and a variety of chemicals, including garden pesticides.

Sewers may dry out an area. According to an Environmental Pollution & Control abstract, Whipple[4] noted that sewers in parts of New Jersey would dry up the streams they were supposed to protect from degradation by septic systems.

14.5.1 U.S. EPA Summary

An excellent summary of land use and environmental impacts of sewers and septic systems has been published.[5] A copy of this summary follows (with references expurgated).

In rural and developing areas, the enforcement of on-site sanitary codes, beginning anywhere from 1945 to the end of the 1960s, has served as a form of land use control. These codes have limited residential development in wetland areas, on soils with a seasonal high water table, including floodplain areas, on steeply sloping areas, and in locations with shallow depth to bedrock because these areas are considered unsuitable for on-site wastewater treatment. Sanitary codes have thus served as a form of de facto zoning, resulting in large lot sizes and a settlement pattern based on suitable soils. The codes have minimized development in some environmentally sensitive areas that would otherwise be unprotected.

Please note that this use of sanitary policy for land use control can have harmful effects. In some states where repair and upgrading of existing systems is considered "new construction," codes have been interpreted to prohibit any upgrading or repair of existing systems. Individual sanitarians have been unwilling to approve repairs or upgrading, to avoid any precedent that might allow further lakeshore development. This not only uses sanitary policy to rule out improvements in sanitation, but forces some residents to think of sewering as the only method that allows community growth. Sanitary and land use policy interact closely, but it is nearly always preferable to consider each openly on its own merits; codes and standards in sanitation should not be used as a crutch to compensate for the absence of goals in land use planning.

The introduction of new forms of wastewater treatment technology that partially or entirely overcome unfavorable site conditions, or that take advantage of more favorable off-site conditions, may enable developers to circumvent these controls. These treatment systems could thus result in significant environmental impacts as a result of the encroachment of housing development on sensitive environmental resources. Also, this could permit a development pattern inconsistent with local goals and objectives. The use of on-site technology such as elevated sand mounds may enable development to occur in areas with a seasonal high water table or shallow depth to bedrock. Off-site treatment such as cluster systems can circumvent on-site limitations altogether and could thus permit development in any of these areas. Impacts from the use of these treatment systems include markedly higher density residential development within existing development areas, a development pattern inconsistent with local goals and objectives, loss of open space buffers between existing developments, and encroachment into environmentally sensitive areas.

To anticipate these impacts, localities should consider conducting land use planning prior to or concurrent with wastewater treatment facilities planning. This would ensure that the suitability of the area for development would be analyzed, that community development goals would be defined, and that appropriate performance standards would be drafted to mitigate impacts of both wastewater treatment facilities construction and associated residential development.

The limited amount of literature available on the land use effects of on-site systems demonstrates the use of sanitary codes to enforce large lot sizes. Local health officials and sanitarians have often become the permitting officials for new housing development and that stipulation has been made for housing densities of 0.5 to 2 dwelling units per acre in order to prevent groundwater pollution.

Generalized dwelling unit per acre zoning in the Seven Rural Lake EIS project areas requires 0.5-acre or larger lots in unsewered areas. Often these lot size requirements have been based on the best professional judgement of sanitarians. These professionals have experienced the need for larger lots because of site limitations or odd lot lines and have recommended larger lots based on the need to protect community health and welfare, not on community development goals.

Alternative on-site technologies may impact lot size requirements. Elevated sand mounds may require larger lots because of larger system areal requirements. Greywater/blackwater separation systems reduce the areal requirements of the soil absorption system.

However, for public health protection, it is unlikely that well separation distances will be reduced. Thus, lot size requirements may not change. Cluster systems featuring centralized collection and off-site treatment will have the same effect on lot size as large-scale centralized collection and treatment systems. When the public health risk from well contamination is avoided, smaller lot sizes are permitted in local zoning codes. For example, Littlefield Township in the Crooked/Pickerel Lakes, Michigan, area allows 4.5 dwelling units to the acre with the provision of public water and sewer. In the Otter Tail Lake, Minnesota, area, provisions for clustered development in the local zoning ordinance allow for 8 to 9 dwelling units to the acre where central sewer service is provided.

The predominant settlement pattern and housing type with standard septic tank/soil absorption systems is reported as single-family detached units in small subdivisions and dispersed low density sprawl patterns. This development pattern has been determined by access to and the spatial distribution of suitable soil. If on-site technologies continue to be used, this development pattern may lead to a situation where the future option to sewer may be precluded because of the great expense of constructing sewers between dispersed houses. Further dependence upon local sanitary codes may thus severely restrict the amount and distribution of developable land in lake areas. Such restrictions may run counter to local growth plans or subdivison plans of large landholders.

One of the most consistent impact findings in the Seven Rural Lake EIS's was that, in the absence of local development controls, centralized collection and treatment systems would induce growth in environmentally sensitive areas such as floodplains, wetlands, and steeply sloping areas. Alternative and innovative forms of wastewater treatment may have similar effects, though to a lesser degree. Historically, sanitary codes have been used as tools to limit or control growth, and as such have become a form of zoning. Some sanitary codes do not permit development of on-site wastewater treatment systems in these marginal areas. However, local municipal officials in many rural lake areas do not have the staff or the budget to conduct land use planning and zoning and do not have formally adopted land use plans. Nor do they have the tools to inventory and analyze their environmental resource base and to formulate performance standards that permit development but prevent significant impacts.

Planning for wastewater treatment facilities gives local municipalities the opportunity to contract for the necessary expertise to conduct land use planning in concurrence with facilities plans. Because the two topics are so closely linked, anticipation of impacts prior to facilities design and formulation of an impact mitigation strategy could save considerable time and expense. An understanding of the environmental resource base, housing types, lot sizes, and existing densities, in conjunction with a program that involves land use planning concurrent with facilities planning, would lead to an environmentally sound wastewater management program.

REFERENCES

1. U.S. EPA. 1980. Innovative and Alternative Technology Assessment Manual. EPA-430/9–78–009.
2. U.S. EPA. 1977. Alternatives for Small Wastewater Treatment Systems, Vol. I. EPA-625/4–77–011.
3. U.S. EPA. 1977. Nationwide Evaluation of Combined Sewer Overflows and Urban Stormwater Discharges. Vol III. EPA-600/2–77–064c.
4. Whipple, Wm. Jr. 1977. Advantages and Disadvantages of Regional Sewerage Systems. Water Resources Research Institute Pub. W 78–00562. Rutgers State University, New Brunswick, NJ.
5. U.S. EPA Region V. 1983. Final-Generic EIS: Wastewater Management in Rural Lake Areas. Water Division, Chicago, IL.

15

The UPC, Uniform Plumbing Code

Codes are written in dogmatic language: "Minimum (area, length, depth, etc.) shall be . . . feet." The reasons for the code commandments are rarely stated, if ever. If the reasons happen to be wrong or obsolete, no one may realize it until after damage has been done.

At first I was apprehensive about employing the UPC[1] (Uniform Plumbing Code) in my line of work. I did not have much choice, as it had been adopted by my jurisdiction and there were no better codes. (The National Plumbing Code was not better. Circa 1978, the Uniform Building Code had a section on septic systems; subsequent editions did not.) Nevertheless, I found out that the UPC has redeeming qualities:

a. It provides uniformity of design and hence a basis for competitive bidding among septic system contractors. This helps keep prices reasonable.

b. It is a time saver. Years ago, I tried to reason with the occasional people who got upset when they found out their unsewered lot was unsuitable for septic systems (the ground was solid rock, for instance, or the land was under water part of the year). I could not satisfy them with 30 to 45 minutes of reasoned explanations about public health hazards, operation of septic systems, etc. One day I hit it right. I said, "You can't have a septic system because it is against the code." To my surprise, this was the end of the argument. Thereafter, I continued using the same technique, with superb results. It seems that most people do not question laws, especially if they don't understand them well enough to argue.

c. It sets only minimum requirements and allows stricter standards. (But the people who enforce its provisions must have the courage to insist on more-than-minimum compliance when necessary, and this is a drawback.)

In the following paragraphs I intend to point out and evaluate important sections of the UPC, and suggest changes if needed.

15.1 PROVISIONS IN THE UNIFORM PLUMBING CODE

Sections 319–321

These sections are very useful to abate nuisances or health hazards. They empower a local jurisdiction to demand abatement of nuisances or hazards even if a septic system was installed and maintained in compliance with previous laws or code editions.

Section 1101

This section requires use of public sewers or a "private sewage disposal system." Connection to public sewers is mandatory except when:

a. The sewer is more than 200 feet away from any building or exterior plumbing.
b. Single-family dwellings and attachments existed before sewer availability, and their approved septic systems work satisfactorily and sewage cannot flow to sewer by gravity.

Section 1102

This section declares that it is unlawful to deposit into any plumbing fixture or device connected to any drainage system " . . . any ashes, cinders, solids, rags, flammable, poisonous or explosive liquids or gases, oils, grease and any other thing whatsoever which would, or could, cause damage to the public sewer, private sewer or private disposal system."

Sodium hydroxide (lye) and sulfuric acid are often used to "declog" a failing leachline or seepage pit. They do not work well (declogging lasts about six months, and repeated treatments are ineffective), and they corrode and weaken the structural integrity of metal and concrete components of septic systems. UPC Section 1102 authorizes prohibition of such practices.

The statement about "poisonous or explosive liquids or gases" fails to take into account that many undesirable compounds can be flushed down in solid form, and that many such substances are not "poisonous" but "toxic." Toxic substances do their damage over a long period of time. Poisonous substances are acutely toxic. Toxic substances include septic system decloggers, such as chlorobenzene, and a variety of chlorinated organic solvents.

I'd like to see the UPC modernized to prohibit deposit of "any ashes, cinders, solids, rags, toxic, corrosive, ignitable, or reactive substances, or any other thing which would, or could, cause damage to the public sewer, private sewer, or private disposal system, or severely degrade groundwater."

The crucial terms "toxic, corrosive, ignitable, or reactive" are defined in the California Administrative Code, Title 22, Division 4. Excerpts are in Appendix G.

Section 1109

Section 1109 establishes authority to require a site-suitability/perk report for use of septic systems.

Section 1119

Section 1119 requires filling abandoned septic tanks, seepage pits, and cesspools (to prevent someone from falling in).

Section I-2

This section contains the statement, "The capacity of any one septic tank and its drainage system shall be limited by the soil structure classification, as specified in Table I-5." This statement needs substantial corrections:

a. Table I-5 refers to soil texture, not soil structure.
b. There is no basis for the requirements in Table I-5. I suppose that its authors were trying to prevent mounding when they developed this table. But, as we have seen in Chapter 13, mounding depends not only on the amount of discharge and the permeability of the soil (which are implicit in Table I-5), but also on the depth of the low-permeability layer, on the hydraulic permeability of this layer, on the depth to groundwater, on the type and geometry of the leachfield, on slope, on drainage, etc. Therefore, I'd change the subject statement to: "The capacity . . . shall be limited by the ability of the leachfield (leachline, seepage pit, seepage bed, hybrid, etc.) to absorb the sewage so that mounding shall not exceed required separations between bottom of leachfield and saturated soil or groundwater table."

An important part of this same section requires that the liquid capacity of a septic tank that serves a dwelling shall be determined by the number of bedrooms (UPC Table I-2). If the structure served is not a dwelling, the capacity is the larger of the values obtained on the bases of:

1. Number of plumbing fixtures (UPC Table 4–1)
2. Actual flow rates

Thus, the septic tank is sized on the basis of the most conservative estimate, and the leachfield absorption area is sized on the basis of septic tank capacity. This conservatism is proper and called for, as peak flows are rarely known.

Section I-3

Various improvements are required in this section.

The minimum amounts for leachfield absorption area requirements are specified to be those in UPC Table I-4. As we saw in previous chapters, this table should not be used in its present form. If perk tests are too expensive or unreliable, the use of something like UPC Table I-4 might be advisable, but not before this table is modified to read as in Table 15.1.

Other leachline requirements found in UPC Section I-3 are carryovers from old traditions: "minimum 150 square feet of trench bottom, and sidewall area in excess of required 12 inches but not to exceed 36 inches below the pipe." Originally, most leachlines were 1.5 to 3 feet wide and had 1 foot of gravel below the pipe. Only the bottom of the leachline was considered to be absorption area. The 150 square feet

Table 15.1 Suggested Modification of Uniform Plumbing Code Table I-4

Class of Soil[a]	Minimum square feet of leaching area per 100 gallons[b] for leachlines or seepage beds serving single homes
1. Sand or gravelly sand; groundwater deeper than 45 feet	55
2. Loamy sand	85
3. Sandy loam, loam	100
4. Silty or clayey loams, uncompacted ("light in weight")	140
5. Clay loams and clays	perk test required

[a]See Glossary for definition of terms.
[b]According to R. Maggard, former IAPMO board president, this refers to "per 100 gallons of septic tank capacity."

was the minimum considered suitable for one bedroom's sewage generation. Also, it provided a storage capacity of about 50 cubic feet within the gravel voids.

Later on, credit was given for extra depth of gravel (i.e., sidewall absorption area) in excess of the first foot below the pipe. For instance, a leachline 3 feet wide with 1 foot of gravel below the pipe is credited with 3 square feet of absorption area per foot of length; with 2 feet of gravel, with 5 square feet; with 3 feet of gravel, with 7 square feet.

I understand that no credit is given for depth of gravel beyond 3 feet below the pipe, because building inspectors' probes go only 3 feet into gravel. The inspectors do not have the time to make a special appointment and be present when the gravel is placed in the leachline trench, and later on they can't verify that the gravel is deeper than 3 feet. This is unfortunate. Sidewall area is far more cost-effective than bottom area, and should be given credit, as is done for seepage pits. Building officials should make use of the "alternate systems" clause (UPC Section I-1h) and find a way to give credit for extra sidewall absorption area. (But beware of wet soil slumping or migrating into the gravel voids after installation. Use geotextile around gravel, or use base material–or sand–below the required 1 to 3 feet of gravel.)

The minimum requirements of 5 feet to groundwater (from bottom of leachlines or leachbeds) and 10 feet to groundwater (from bottom of seepage pits) are adequate only if mounding is not a problem and if soils are not excessively permeable (less than 5 mpi) and do not have channels or fractures.

Section I-6e

Per UPC, where two or more drain lines (from septic tank to leachlines, pits, etc.) are installed, a distribution box shall be installed to ensure equal flow.

As we saw in Chapter 10, Section 10.2, a distribution box does not split the flow equally, unless the flow is massive.

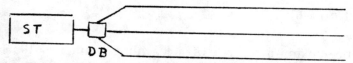

Figure 15.1 Top view of septic system with a distribution box. ST = septic tank and DB = distribution box.

Figure 15.2 Septic system without a distribution box.

Septic systems for single houses should not rely on equal splitting. The old-fashioned methods should be revived.

Today, a septic system with leachlines on level ground may appear as in Figure 15.1. But a long time ago, before distribution boxes were introduced, the same septic system would have appeared as in Figure 15.2. The old-fashioned design has a perpendicular leachline branch at the end to connect all other branches. This way, one branch cannot be full and failing while another one is dry. This same design should be mandatory when leachlines are on level ground.

Section I-6h

A dosing tank is required if a septic tank discharges to leachlines with lengths adding up to more than 500 feet. The massive surge from the dosing tank helps distribute the flow to the end of the leachline branches.

Dosing tanks do malfunction. They leak and do not release effluent in a surge. A provision for inspecting and maintaining dosing tanks might do much good. (An electrical or mechanical alarm would help.)

Section I-6i

The maximum length of a leachline branch is 100 feet. Winneberger states that this requirement is a carryover from the times when leachpipes were made of clay and broke when vehicles ran over the leachlines.[2]

As I see it, sewage tends to be discharged near the beginning of the leachpipe. If the bottom of the leachline trench sloped down 2% to 4% toward its end, there would be no good reason to restrict maximum length to 100 feet. The sewage would spread through the leachline, accumulate near the end, and move up and down as in a seepage pit. If this vertical movement of the surface of the ponded sewage amounts to 1 foot in a 2% sloped bottom, this would wet/dry 50 feet of bottom; and this would help with aeration.

The required grade of "lines" is "level to 3 inches/100 feet." This has been interpreted to mean that leachpipes and bottom area must be nearly level. But it is much better to interpret it as referring only to lines, i.e., pipes. The problem is that some leachlines with sloped bottoms might have more than 3 feet of gravel at their end, and this might frustrate building inspectors who have to check the depth of gravel.

The minimum spacing between leachline trenches, side-to-side, is 4 feet, plus 2 feet for each foot of gravel below the leachpipe in excess of the first foot. Fair enough.

Section I-7

Seepage pits are required to be circular and to have a lining of bricks or blocks or "other approved materials." Where gravel is cheap, it may be cost-effective to place a perforated pipe, 4 inches in diameter, vertically into a seepage pit excavation, and fill the excavation with gravel. The effluent that is not absorbed at the top of the column of gravel would fall into the vertical pipe and be absorbed below. (A clogging mat can form over the top of the gravel if toilet paper, napkins, and similar materials escape the septic tank. Soil migra-

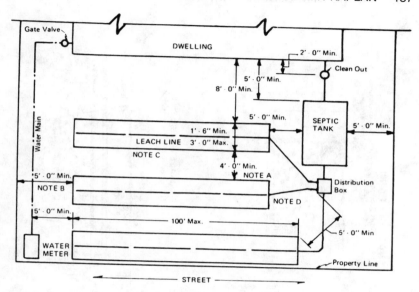

LEACH LINE SYSTEM

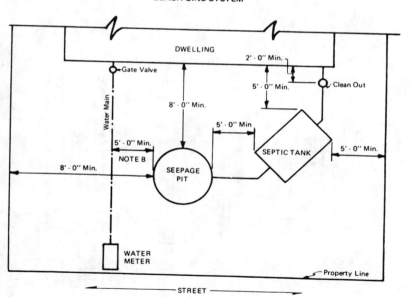

SEEPAGE PIT SYSTEM

Figure 15.3 UPC design requirements for leachlines and seepage pit. (Source: R. Maggard.)

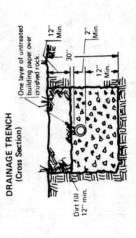

DRAINAGE TRENCH
(Cross Section)

One layer of untreated building paper over crushed rock.

12" Min.

2" Min.

30"

12" Min.

Dirt fill 12" min.

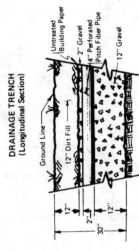

DRAINAGE TRENCH
(Longitudinal Section)

Ground Line

Untreated Building Paper

2" Gravel

4" Perforated Pitch Fiber Pipe

12" Gravel

12" Dirt Fill

12"

2"

30"

12"

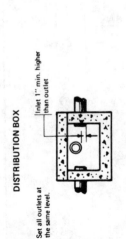

DISTRIBUTION BOX

Inlet 1'' min. higher than outlet

Set all outlets at the same level.

SEPTIC TANK SIZE

BEDROOMS	GALLONS
1 or 2	750
3	1,000
4	1,200
5 or 6	1,500

NOTE A:
Add two (2) feet to this dimension for each additional foot of gravel below the twelve (12) inch gravel bed in trench.
NOTE B:
Where no water main exists, the leach line or seepage pit may be located a minimum of five (5) feet from side property line.
NOTE C:
Leach line must contain at least one hundred and fifty (150) square feet of trench bottom. There must be sufficient yard space to increase the leach line by one hundred (100) percent.
NOTE D:
These lines from the distribution box to the leach line area shall be water - tight lines.

Figure 15.4 UPC requirements for septic tank size, distribution boxes, and drainage trenches (leachlines). (Source: R. Maggard.)

Minimum Horizontal Distance in Clear Required From:	Bldg. Sewer	Septic Tank	Disposal Field	Seepage Pit or Cesspool
Buildings or Structures [1]	2 ft.	5 ft.	8 ft.	8 ft.
Property line adjoining private property	Clear	5 ft.	5 ft.	8 ft.
Water supply wells	50 ft. [2]	50 ft.	100 ft.	150 ft.
Streams and lakes	50 ft.	50 ft.	50 ft.	100 ft.
Large trees	–	10 ft.	–	10 ft.
Seepage pits or cesspools	–	5 ft.	5 ft.	12 ft.
Disposal field	–	5 ft.	4 ft.	5 ft.
Domestic water line	1 ft.	5 ft.	5 ft.	5 ft.
Distribution box	–	–	5 ft.	5 ft.

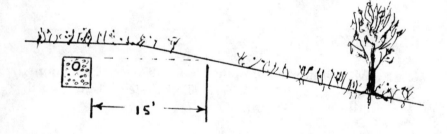

NOTES: When disposal fields and/or seepage pits are installed in sloping ground, the minimum horizontal distance between any part of the leaching system and ground surface shall be fifteen (15) feet.

1. Including porches and steps whether covered or uncovered, breezeways, roofed porte · cocheres, roofed patios, car ports, covered walks, covered driveways and similar structures or appurtenances.

2. All non-metallic drainage piping shall clear domestic water supply wells by at least fifty (50) feet. This distance may be reduced to not less than twenty-five (25) feet when approved type metallic piping is installed.

Where special hazards are involved, the distance required shall be increased, as may be directed by the Health Officer or the Administrative Authority.

Figure 15.5 UPC requirements for sewage system location. (Source: R. Maggard.)

tion or slumping can plug portions of the gravel volume.) Pits full of gravel are not likely to be a lethal hazard after they are abandoned and forgotten.

UPC design requirements have been summarized in Figures 15.3, 15.4, and 15.5. Septic system designers should note that:

1. Leachlines may be installed right next to the property line adjacent to a street, but (any portion of a leachline) must be 5 feet from any other property line.
2. The UPC does not require that leachlines be straight. They can be curved at 90 degree angles. This is not a good practice where mounding may occur.

For many years, there has been talk of the Uniform Plumbing Code being incorporated into the Uniform Building Code. It appears that this is finally about to happen. I hope that the Uniform Plumbing Code will be improved before this event occurs.

REFERENCES

1. IAPMO. Uniform Plumbing Code, 1976 to 1988 editions.
2. Winneberger, J.T. 1984. *Septic Tank Systems*. Butterworth Publishers, Stoneham, MA.

16

Ethics

Various ethical considerations come into play in any governmental activity. Occasionally, some ethical standards conflict with other ethical standards. What follows is not a lesson in ethics. It is a sample of illustrations to whet the reader's appetite for more. And much more is presented in Appendices Q, R, and S and particularly in Appendix R. Whoever becomes a "septic system specialist" in a governmental entity should be prepared to face similar quandaries and dilemmas and choose his/her own course of action. An excellent guide to ethical courses of action has been written by Gunn and Vesilind.[1]

16.1 LAND USE

Fairness is part of ethics.

Early in this century, a certain county allowed minuscule lot sizes (2,500 square feet) in subdivisions with no sewers. In modern times, to be fair to the owners of such lots who could not build (for lack of space for leachlines), the county's board of supervisors allowed the use of holding tanks,* under permit to Environmental Health and local sanitation districts. This is what happened in the ensuing years:

a. Many holding tank owners shot holes in the tanks after they were installed to induce leakage and save on pumping costs.
b. On occasion, holding tank pumpings were dumped near creeks.
c. Some people merged three or five small lots and built expensive homes with septic tank systems. Later on, their small-lot neighbors built 400-square-foot "houses," with holding tanks, right next to the expensive homes and affected property values.

It is difficult to be fair to all.

16.2 RESTRICTIVE STANDARDS

In the late 1970s, I faced an ethical dilemma. I reviewed our local standards for determining site suitability. I found that, in general, our standards restricting maximum slopes to 25% and 30% had no factual basis.** If I relaxed the standards, major areas in our scenic mountains would be opened for development. Development on steep slopes could cause substantial environmental impacts.

---

*A holding tank holds all the effluent from a home, so that (in theory) none of it is discharged into the ground or nearby streams. It is made of coated steel, looks like a septic tank (but has one manhole with a removable lid at ground level), holds 1,500 or 2,000 gallons, and has an electrical or mechanical alarm that signals when it is 2/3 full. At a per capita discharge of 50 gallons/day, a family fills it up in about 10 days. It must then be pumped at a cost of $80–100 per 1,000 gallons, say $3,000 or $4,000 per year under daily use.

**These standards followed state guidelines. They were at odds with principles decribed in Chapter 5 and 6: the steeper the slope, the lower the danger of sewage surfacing. However, these standards are proper in some areas that have unstable (landslide-prone) slopes.

I had two choices. One was being ethical in the sense of being honest, and eliminating an unwarranted restriction (i.e., being fair to owners of steep lots). The other one was being ethical with regard to preservation of nature and the environment: I would keep quiet about a restriction I thought unwarranted, and wait for a very long time or forever until a knowledgeable private consultant challenged the standard (none had as yet).

Being ethical in one way meant being unethical in the other way.

After much thought, I decided that, as a private citizen, I was entitled to try to preserve nature. But as a county employee, my responsibility was to enable development where it was feasible. So I relaxed the standards. I might have acted differently if I had not had confidence in county planning agencies.

With some misgivings, I have done the same thing with regard to another unwarranted restriction: maximum allowable perk time equals 60 mpi. It is possible to have a septic system in a 400-mpi soil. My misgivings are due to fears that the relaxed standards might be copied by other jurisdictions that will not require (or that will not know how to require) careful design, installation, and use of septic systems in high-mpi soils.

16.3 DISCIPLINARY ACTIONS AGAINST PERK CONSULTANTS

It is not easy for me to come down hard on a nice perk consultant I have caught cheating, especially if we have been on friendly terms. But things that must be done, must be done. My feelings about dishonest or incompetent perk consultants are strong.

As I see it, it is not ethical when a mugger steals $100 at gunpoint. He gains $100 and his victim loses $100. This is fair mugging, as mugging goes. It is far worse when a perk consultant makes $400 or $1,000 with a spuriously favorable site-suitability/perk report that burdens his client (or the person who buys a worthless lot from his client) with a $10,000 or $80,000 loss. Keeping this in mind, and remembering the distress of some victims I met, I have had no qualms when I went after consultants who should not be in business.

Disciplining involves ethical and technical judgments. Technically, I have had to figure out if a consultant's "mistake" was due to mere ignorance, to incompetence, or to outright cheating. Ethically, I have tried to take a course of action that first and foremost protects the public from the consultant, and that also does not harm the consultant more than necessary to ensure that "mistakes" do not occur again.

Of course, I am not the final judge or arbiter. More about this, and about legal procedures involved in disciplinary actions, is described in the next chapter.

REFERENCE

1.   Gunn, A.S., and Vesilind, P.A. 1986. *Environmental Ethics for Engineers*. Lewis Publishers, Chelsea, MI.

17

Work Privileges and Certification of Perk Consultants

At the California Environmental Health 1985 convention, M. Vinatieri* stated that in his county about 600 septic systems are designed by perk consultants and installed annually, and that 30% fail after two to three years. His words are echoed through much of the nation, as few institutions train and certify or license perk consultants.

I doubt that many surgeons dare practice without thorough grounding in anatomy. But every month or two I meet a prospective consultant who insists he can perform site evaluations and septic system designs although he has received no training in the subject. Professors of civil engineering have told me they devote "about two hours" to septic systems in their lectures. It seems that scant instruction has convinced a good number of engineers I have met that they are quite capable of measuring how fast water is absorbed in a test hole, looking up a table, and writing a report—since, according to them, that's all there is to practicing as a perk consultant. They insist that the law allows registered civil engineers to practice in the field of septic systems. My pointing out that they must be competent**

---

*Director of Environmental Health, Sonoma County.
**One of the fundamental canons of ethics of the American Society of Civil Engineers is embodied in the California Administrative Code, Title 16, Section 415: "A professional engineer . . . shall practice only in the field or fields in which he is . . . fully competent and proficient."

often fails to persuade them that their "right to practice" is limited by the law.

I understand that most of the complaints filed with the Board of Registration for Professional Engineers (California Department of Consumer Affairs) relate to septic systems, and that, therefore, only soil engineers will be licensed to practice as "perk consultants" starting in 1987.* But I doubt that this measure will have a significant effect. There is a specific body of knowledge that must be learned. This knowledge being absent, no license or amount of "experience" will result in satisfactory performance. (The field is full of consultants who have many years of experience doing the wrong things.)

Ignorance is perhaps the most common problem evident in site-evaluation/perk reports for sites with unusual conditions, but not necessarily the most serious problem. Dishonesty is.

Dishonesty or "cheating" may range from ingenious trickery (for instance, secretly drilling a leachline perk test hole 10 feet down into a low-permeability clay, filling it with sand to 5 feet, placing 2 inches of gravel over this sand, and "proving" that the clay's absorption rate is satisfactory) to falsification of facts (for instance, failing to see and report groundwater at 2 feet when excavating a 10-foot-deep trench in a swamp).

For many years, the California Directors of Environmental Health have pondered how to solve the above-mentioned problems. In San Bernardino County, we have taken substantial steps in this regard. An account follows.

In the 1970s, in southern California, an engineering firm specialized in producing satisfactory perk reports with false data. They never "noticed" any problems with shallow groundwater, solid rock, or steep slopes. I reviewed some of their amazing reports, and so did a colleague in a neighboring county environmental health unit, William Leuer, P.E.

---

*Note: A young engineer who could not qualify sued the state and won. Currently, the limitation to soil engineers does not apply.

Both of us refused to approve perk reports prepared by the subject firm. Leuer notified the Registration Board.

Eventually, these events gave impetus to an ordinance which defined who qualified as a perk consultant, required demonstration of competence, and established causes and procedures for revocation of perk testing privileges.

It is important to emphasize the word "privileges." County counsel determined that, since the county (Environmental Health) requires, regulates, and establishes the rules for perk report submittal, it is also empowered to establish who may and who may not prepare such reports.

Many engineers have told me our ordinance is not legal, and I have heard the same from sanitarians in other counties who believe that registered engineers with unrevoked licenses cannot be prevented from practicing in the field of septic systems even if they are ignorant and crooked. Maybe so, maybe not. Our ordinance has been at work, and it has worked very well. For reasons that will become clear in the paragraphs below, no one has challenged it in court. The pertinent portions of this ordinance and of the hearing procedure for suspension or revocation of perk testing privileges are in Appendix H.

After the ordinance was adopted, as part of a training program for prospective perk consultants, I prepared a very friendly written examination designed to give anyone ample opportunities to show his competence. It could not be friendlier:

a. I suggested reading materials for the examination.
b. The examination was open-book, and I was there to answer any questions at any time (except for direct answers to examination questions).
c. Although the examination took 2.5 to 3.5 hours to complete, one could work on it the whole day.
d. The date and time of the examination were pretty much up to the examinee to decide.
e. Any answer that made sense received full credit.

Individuals who had been submitting satisfactory perk reports for years were "grandfathered." All others had to pass the exam, or proceed at their own risk.

The outcome has been very revealing. About half of the prospective perk testers (or consultants) who inquired about practicing within our jurisdiction changed their minds when they heard they would be examined (as described above), even though they would be given the opportunity to challenge the outcome in front of the Registration Board. Of the other half, one-third received a score of 88%-89%; the rest had scores below 55%. The minimum passing score was 70%. Most found the test fair. One of the high and one of the low scorers found the test very easy. Not one among those who failed challenged the examination scores: all of them realized that they were not as competent as they thought, and that they could get into trouble if they practiced in our county.

It is of interest to note that the high scorers took the trouble to read the literature that was suggested to them (parts of the EPA Design Manual, Winneberger's *Septic Tank Systems*, the Uniform Plumbing Code, and the local Procedures Manual). The low scorers did not. Another note of interest is that the two who protested the most about having to be examined performed the worst.*

Training and testing serve to "keep away" consultants who

---

*One of them was a consultant who had been "perking" in a neighboring county. Under our local requirements, perk report preparers (i.e., consultants) must certify that their septic system design will not cause groundwater degradation. So the exam had a multiple-choice question requiring the identification of the most common natural septic system pollutant. (Regional newspapers had published articles on it, so one did not have to be an expert to choose the right answer.) When he encountered the subject question in the exam, he looked at me and protested furiously, "This is chemistry!" His score turned out to be about 22%.

The other one was unforgettable. Our conversation was more or less as follows (he is E; I am K):

E: I do not need to take an exam.

K: You may know a lot, but county code requires that you demonstrate your knowledge.

E: . . . [His answer was a long recital of qualifications as owner of (and principal engineer in) an established firm—with no mention of accomplishments in regard to septic systems.]

K: So what. That doesn't demonstrate knowledge about septic systems.

are less than fully competent. The site evaluation or perk report requirements described in Appendix I provide other mechanisms to check up on competence and honesty.

From all the above, it appears that local jurisdictions can and should play an active role in helping professional registration boards protect the public and the interests of competent professionals.

Note: The examination of prospective perk testers was discontinued in 1987, before the State Attorney General determined that a county was entitled to test the engineers' knowledge *regarding local perk testing procedures.*

E: It is illegal for you to require an examination.

K: Fine. Sue the county.

He decided to take the exam at once, without any preparation, and without reading any of the suggested literature.

His score turned out to be about 24%.

18

Standard Site Suitability/Perk Report

The previous chapter illustrates why prospective perk consultants should be trained and/or examined and certified to ensure that they know at least a bare minimum about perk topics. In this chapter we will explore a way to ensure that consultants who do not measure up in other respects can be persuaded to change their line of business. The means to this end is the "site suitability/soil percolation test report."

"Site suitability" and "soils report" are terms used in completely different contexts by planning and building departments. Furthermore, it is awkward to use a full name with 17 syllables. Even the often used name, "soils percolation test report," is too long. Hence, we'll call it "perk report."

Under the auspices of Riverside and San Bernardino Counties, in the mid-1970s about a dozen southern California engineers formed an "Ad Hoc Percolation Test Committee" to try to standardize the perk requirements of both counties. After more than 400 man-hours were devoted to the task, in 1979 the committee produced a document which was useful as a framework. But due to unmet concerns, each of the two counties modified this framework to satisfy its own needs. Uniform requirements were still elusive.

Between 1979 and 1984 it became obvious that, for perk report preparers to be held accountable, and subject to suspension or revocation of "perk consulting" privileges, the perk standards had to be specific and detailed. Anyone can make a mistake, and "forget" about noticing solid rock at 2 feet depth. But if local perk standards require noticing such a constraint, excuses are not plausible.

Also, the Professional Registration Boards use a local jurisdiction's perk standards when determining whether a perk consultant has done something wrong. If the standards are vague, a Board can do little or nothing.

In order to overcome resistance from some local perk consultants who did not want to be subjected to clear and specific requirements, I waited until some of them made "mistakes." As they did, I "slipped in" pertinent requirements in our county "perk standards manual," and encountered no opposition.

A partial list of these "mistakes" follows:

1. Signed report prepared by unqualified technicians.
2. Forged data.
3. Topography was ignored in the design: part of the leach-line was 16 feet below ground, and part of it was up in the air.
4. Conducted perk tests without notifying the county "septic systems" specialist; so, results could not be verified years after testing.
5. Tested the wrong lot; the "right" lot would have passed.
6. Tested the wrong lot; the "right" lot would not have passed.
7. Did not find caliche while boring with a screw-type auger through caliche strata.
8. Did not find groundwater (in the dry season).
9. Did not know that water level moves faster in a gravel-packed test hole.
10. Soil over solid bedrock was too thin for leachlines; so, he recommended seepage pits.
11. Placed leachlines over stratified soils next to a road cut. The strata sloped downward toward the cut.

In 1985 a reconstituted "Ad Hoc Perk Committee" tried again to make the perk standards uniform in the two counties. The result is that, after many meetings, it suggested adoption of the San Bernardino County format with "standards" that reflected the wishes of the perk consultants. (These included: low absorption area requirements, measurement precision of $1/4$ inch for leachline perk tests, and continued use of $Q=9FD/Lt$ formula.)

I understand that in one of the committee meetings a very experienced septic systems contractor pleaded with members of the committee not to reduce absorption area requirements below the EPA's recommendations; he himself had found out the hard way that lower requirements do not work well. If the requirements were lowered, he would still install larger-than-required absorption areas, and lose business to less knowledgeable or less scrupulous contractors. But he was overruled because most members of the committee "did not know of any failures" with reduced absorption areas, and as they put it, "it is easier to fit a smaller leachfield in a small lot."

On the positive side, I learned two things while working with and against this committee:

1. It pays to "keep in touch" with perk consultants. There are things one does not think about when developing standards. For instance, at one meeting I proposed that the standard diameter of leachline test holes should be 6 inches. I managed to come out alive, with the distinct impression that the perk consultants were not agreeable to my suggestion. (They had already bought equipment to bore 8- to 10-inch-diameter holes.)
2. Private perk consultants alone should not be allowed to set standards. As one of the consultants forcefully stated during a heated discussion, his concern and responsibility were to his own client. It became very clear during some of the perk meetings that the concerns of some of the consultants had little to do with environmental health.

My Riverside County environmental health colleagues* and I met and evaluated the Committee's latest proposals.

---

*Wm. Leuer, P.E., and R. Luchs, R.S.

We were (practically) of one mind as to what the perk report standards should be. They are shown in Appendix I.

19

Strategic Considerations About the Use of Septic Systems

From the point of view of a governmental jurisdiction, various strategies for use of septic systems could be considered.

Ideally, the jurisdiction's land use planners have determined which areas are to be developed with and without sewers, where the discharges will go, the amount of these discharges, and the carrying capacity of each area for the use of septic systems (among other things).

It is costly and ineffective to require piecemeal studies of environmental impacts as growth occurs; often, such studies are commissioned only after growth has overwhelmed an area's carrying capacity.

Since the amount of leachfield absorption area is fairly inelastic in respect to perk times (1- to 60-mpi range), it might be cost-effective to require a conservative size of leachfield in communities with no (or moderate) soil and site problems and dispense with costly perk reports. For instance, for single homes with 100% or 200% reserve area for replacements, these conservative sizes could be 70 square feet (very sandy soils) and 140 square feet (other soils with little clay) of leachline absorption area per hundred gallons of septic tank capacity—and half as much absorption area for seepage pits. Anyone would be free to hire a perk consultant, file a perk report, and reduce what might appear to be an excessive requirement.

As we saw in Chapters 8 and 17, the best strategy for

ensuring long-lived, low-cost septic systems must take into account all of the factors that contribute to premature failure:

a. Perk consultants should be knowledgeable, and licensed or certified.
b. Septic system contractors should be given at least a minimum amount of instruction as to what the soil and site conditions are in their localities, what hazards to avoid, and when to notify or consult with county specialists. A bare-bones certification program might help, too.
c. Septic system users should be given a short, readable leaflet about the use and care of septic systems. (One such leaflet is shown in Appendix J.) If politically viable, they should be licensed to use septic systems after demonstrating they are at least aware of where their sewage goes to. (I have met people who didn't know that their houses were on septic systems. Some wondered why they never got a sewer bill.)

Many people buy houses served by septic systems without giving too much thought to what a septic system is like. After their septic systems fail, they call their local jurisdiction and ask, "What kind of septic system do I have? Where is it located?" Well, in the very near future, cheap, compact data storage in optical (laser) disks may permit easy and fast retrieval of plot plans showing the layout of each and every septic system in a jurisdiction.

On the basis of money alone, septic sytems and their users deserve a good break. At about $1,000 per leachfield, the total investment in leachfields in San Bernardino county is about $100,000,000; in California, about $2,000,000,000; in the United States, about $20,000,000,000. Since at present the average leachfield lasts roughly 10 years, a jurisdiction's program that merely increases average life of leachfields by 10% might have a superb cost/benefit ratio.

20

Concluding Thoughts

After a new house was built, the villagers released chickens near it, and observed their movements. Where they first saw a chicken "leaving its mark" on the ground, that's where they built a [20-foot-deep] cesspool for the new house. And their cesspools lasted forever.

> —Adapted from a tale of anonymous authorship. (Note: the village's ground was flat and made of dry, permeable sand all the way down to 300 feet.)

As far as I am concerned, if it works, don't knock it. If site and soil conditions are good, any chicken can do as well as a competent perk consultant, and for much less money.

But if site and soil conditions are not good, the difference between chickens and experts may become quite obvious after a while. For instance, I have just visited a large sewage disposal site for a residential development in another county. Black septage is oozing out of the soil at the lower portions of a concave 10-acre basin, and is forming a mosquito-laden swamp. The septic system was designed by a major engineering company, which tested the upper 2 feet of a thin (8-foot) soil mantle over impermeable bedrock, and concluded that the site was quite suitable for leachlines.

Methodologies currently in use in some jurisdictions, like that in the anonymous-authorship tale, may work under favorable local soil conditions. But they cannot be applied blindly where the conditions are different.

In practice, one is likely to encounter infinitely varied conditions or problems. A cookbook approach to site evaluation or design of septic systems is nearly worthless when the problems are of an unusual variety, as is often the case. Therefore, in this book, I have attempted to explore and explain reasons, causes, and effects.* These are the key to solving problems. Additional in-depth perspectives may be found in Appendixes M through S.

The reader might wish to test his understanding of the subject matter by trying to answer the questions provided in Appendix L. Questions like these might appear in certification or competency examinations.

I hope that any such examinations will be open-book. Understanding, not memorization, is what makes the difference between a competent and a not-too-competent perk consultant.

---

*While doing so, I might have exposed my own biases or misconceptions. If any of my readers can teach me something, I'd be most appreciative. To this end, feel free to write to me, P.O. Box 522, Calimesa, CA 92320.

Appendix A

ORGANISMS IN SEWAGE

Table A.1 Human Enteric Viruses in Sewage

Virus	Number of Types	Diseases Caused
Enteroviruses:		
Poliovirus	3	Meningitis, paralysis, fever
Echovirus	31	Meningitis, diarrhea, rash, fever, respiratory disease
Coxsackie virus	23	Meningitis, herpangina, fever, respiratory disease
Coxsackie virus	6	Myocarditis, congenital heart anomalies, pleurodynia, respiratory disease, fever, rash, meningitis
New enteroviruses (Types 68-71)	4	Meningitis, encephalitis, acute hemorrhagic conjunctivitis, fever, respiratory disease
Hepatitis Type A (enterovirus 72?)	4	Infectious hepatitis
Norwalk virus	1	Diarrhea, vomiting, fever
Calicivirus	1	Gastroenteritis
Astrovirus	1	Gastroenteritis
Reovirus	3	Not clearly established
Rotavirus	2	Diarrhea, vomiting
Adenovirus	37	Respiratory disease, eye infections

Source: EPA-625/1-83-016.

Table A.2   Bacteria and Parasites in Sewage and Sludge

Group	Pathogen	Disease Caused
Bacteria	Salmonella (1700 types)	Typhoid, paratyphoid, salmonellosis
	Shigella	Bacillary dysentery
	Enteropathogenic *Escherichia coli*	Gastroenteritis
	Yersinia enterocolitica	Gastroenteritis
	Campylobacter jejuni	Gastroenteritis
	Vibrio cholerae	Cholera
	Leptospira	Weil's disease
Protozoa	*Entamoeba histolytica*	Amebic dysentery, liver abscess, colonic ulceration
	Giardia lamblia	Diarrhea, malabsorption
	Balantidium coli	Mild diarrhea, colonic ulceration
Helminths	*Ascaris lumbricoides* (Roundworm)	Ascariasis
	Ancyclostoma duodenale (Hookworm)	Anemia
	Necator americanus (Hookworm)	Anemia
	Taenia saginata (Tapeworm)	Taeniasis

Source: EPA-625/1-83-016.

Appendix B

2.2 "FAILURE" OF LEACHFIELDS

Leachfields can be built to last indefinitely. If leachfields do "fail," it is not the *leachfields'* failure.

Perhaps the myth about leachfield failures came about for a practical reason.

The first or second year I worked as a sanitarian, I was asked to participate in a sanitary survey of a target area. The procedure was simple. We had to go house to house and note any indications of leachfield failure: sewage on top of the ground, odor, drainage hose or pipes sticking out of the house, etc. And, when possible, we had to interview the residents and ask about "failure." If 2% of the homes had "failures," the area was declared a "failure problem area," and had to be sewered.

The procedure seemed illogical. If we conducted a survey of "failing cars" and added up all the "failing cars" in a neighborhood, we would be counting cars that had conked out the day of the survey along with cars that had been lying around since Methuselah's time. When we add up prevalence and incidence, we get a number that means nothing. Then, we would have to go tell the people in the neighborhood that since x% of their cars have "failed," the whole neighborhood must find an alternative means of transportation! Yet this is what the sanitary survey was about.

To make matters worse, I had asked, "How long is a good leachfield supposed to last?" The answer was, "About 20 years." Well, assuming that failures were to occur randomly in time and leachfields last 20 years, at any given year 5% of the leachfields would fail! And 5% is higher than the critical 2% standard. I voiced my concerns, but received the type of stares reserved for the naive.

Months later, I learned that if the health officer declares an area a "failure problem area," state and federal funds become available for sewering it, and cover 87.5% of the total cost. A real bargain. (So, I had to admit to myself that something absolutely illogical could be perfectly reasonable.) But nowadays, state and federal funds have dried up.

2.2 FLUCTUATION OF SEWAGE LEVELS IN SEEPAGE PITS AND LEACHLINES

The Uniform Plumbing Code requires that a leachline have at least 150 ft^2 of bottom area, and that seepage pits be 4–6 ft in diameter.

The volume of water in a trench = width × length × height

or

$$V = W \times L \times H$$

And the volume of water in a cylindrical pit is

$$V' = \pi r^2 H'$$

Taking derivatives, we have that

$$dV/dH = W \times L$$

and

$$dV'/dH' = \pi r^2$$

Hence, rearranging and combining,

$$dH'/dH = W \times L/\pi r^2$$

Since $W \times L = 150$ ft^2, and r is 2–3 ft, we can substitute these values in the formula above to get

$$150/\pi(2^2) = 12$$

and

$$150/\pi(3^2) = 5.3$$

So the level of sewage fluctuates 5.3 to 12 times more in a pit than in an open minimum-size leachline trench. But after the leachline trench is filled with gravel, the fluctuation in the leachline trench must increase by a factor of about 3, because gravel voids are about 1/3. Therefore, the sewage fluctuation difference is at least 1.8 to 4 times.

Appendix C

5.1 COMMENT

In the formula $H = h + h'$, the surface head h is also called "pressure head," and h' is called "gravitational head" or "soil head."

5.5.1 THE UNWANTED IRRIGATION CANAL

Table C.1 Percent Slope and Value of the Sine[a]

Slope (%)	Sine	Slope (%)	Sine
1.000	0.010	23.414	0.228
2.000	0.020	24.472	0.238
3.001	0.030	25.534	0.247
4.002	0.040	26.602	0.257
5.004	0.050	27.676	0.267
6.007	0.060	28.755	0.276
7.011	0.070	29.841	0.286
8.017	0.080	30.934	0.296
9.024	0.090	32.033	0.305
10.033	0.100	33.139	0.315
11.045	0.110	34.252	0.324
12.058	0.120	35.374	0.333
13.074	0.130	36.503	0.343
14.092	0.140	37.640	0.352
15.114	0.149	38.786	0.362
16.138	0.159	39.941	0.371
17.166	0.169	41.105	0.380
18.197	0.179	42.279	0.389
19.232	0.189	43.463	0.399
20.271	0.199	44.657	0.408
21.314	0.208	45.862	0.417
22.362	0.218	47.078	0.426

[a]Interpolate when necessary.

5.5.2 MEASURING k

Let us visualize a 4-in. layer of clay over a 4-in. layer of sand. If we measure the vertical k's (saturated flow) of the clay and of the sand, we might get, say, 0.001 ft/day and 1 ft/day, respectively. Under a leachfield, water movement is usually not under pressure. The water moving through the clay will encounter an unsaturated sand with lower permeability than that measured under saturated flow. In the lab, with an h of 30 ft, so much water is transmitted through the clay that the sand below becomes more saturated, and more permeable. The joint average permeability of the clay and the sand strata is overestimated. This is fine when the test is for liners, but not when it is for leachfield design.

The same type of problem affects the harmonic-mean k formula.

5.5.3 WETTING FRONTS

The round figure to the left represents the clay soil. Capillary suction is strong, in all directions; movement downward is slow. Most of the movement within the sand is downward and fast.

5.5.4 EVAPORATION

As shown in Table 5.1, 33% of 4 ft, or 1.33 ft, will be lost to evaporation per year. Multiplying 1.33 ft by (3 × 100) ft² of evaporative surface equals 400 ft³ lost to evaporation per year. And 300 gal of sewage/day equals 300 × 360/7.5 = 14,400 ft³ discharged.

So, 400/14,400 = .027, or 3%.

5.5.5 BALANCE

Clay will suck up water from the silt, as it has more smaller-diameter capillaries than the silt, and the silt has more (and more continuous) smaller-diameter capillaries than the sand. The balance will tilt down toward the clay.

5.5.6 SEEPAGE

Condition (a) (clay overlying coarse sand) will.

Appendix D

6.3 GEOMETRIC MEAN VERSUS ARITHMETIC MEAN

Perhaps another mental experiment will make it easier to visualize why the arithmetic mean is appropriate while the geometric mean is not.

Let us assume that we have a large barrel of water. Let us divide its 300-in.^{2} bottom into three sections, 100 in.^{2} each. Let's put one hole in each of these sections. The areas of the holes are 1, 10, and 100 in.^{2}; the flows through the holes are proportional to their areas, say, 1 in.^{3}/sec/in.^{2} of hole at a given head. The total amount of infiltration or flow through the barrel's bottom will be $1 + 10 + 100 = 111$ in.^{3}/sec.

Now, let us see if the geometric mean of the infiltration rates per 100 in.^{2} will give us the total amount of flow after it is multiplied by the total area of the bottom. The cubic root of $(1/100) \times (10/100) \times (100/100)$ is the geometric mean, and it is equal to 10/100 (10 in.^{3}/sec/100 in.^{2} of barrel bottom). If we multiply this mean by the total area of the bottom, 300 in.^{2}, we get a total flow rate of 30 in.^{3}/sec, which is quite a bit less than 111.

The arithmetic mean is equal to 1/3 of $(1/100 + 10/100 + 100/100)$ or 111/300; and this, multiplied by the total area of the bottom, 300 in.^{2}, gives 111 in.^{3}/sec, which is correct.

6.4.1 GRAVEL PACKING CORRECTION FACTORS

See Table D.1, starting on the next page. Multiply your measurements in ipm (or mpi) by the factors given to obtain ipm (or mpi) in a hole with no gravel packing. P = porosity of gravel; C = ratio of diameters, hole/pipe (or ratio of outer over inner diameter of gravel pack).

Example: You measure 10 mpi. Porosity is 0.30 and C is 1.6. On next page, under C = 1.6, look down the P column until you find 0.30. Move horizontally to the "× mpi" column and you will see 2.77. Multiply 10 mpi times 2.77 and the result is 28 mpi.

6.4.2 ACCURACY OF MEASUREMENT WITH NONVERTICAL TAPE

The nonvertical tape or rod will measure a longer distance equal to a in the triangle below.

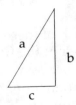

The correct distance is b. So the error is a – b. Using the theorem of Pythagoras, we can calculate b and subtract it from a.

In the first problem, a = 4.5 × 12 = 54 in.,

$$b = \sqrt{(a^2 - c^2)} = 53.66 \text{ in.,}$$

and the error is a – b = 0.34 in. [5/16 in.]

In the second problem, the same procedure is followed. The error is 6/16 in.

Table D.1   Gravel Packing Correction Factors

C = 1.2

P	× ipm	× mpi
.24	.3057439	3.270711
.26	.3240138	3.086288
.28	.3422837	2.921553
.30	.3605536	2.773513
.32	.3788236	2.639751
.34	.3970934	2.518299
.36	.4153633	2.407531
.3800001	.4336333	2.306096
.4000001	.4519032	2.212863
.4200001	.4701731	2.126876
.4400001	.488443	2.047322
.4600001	.5067129	1.973504
.4800001	.5249828	1.904824
.5000001	.5432527	1.840764

C = 1.4

P	× ipm	× mpi
.24	.3057439	3.270711
.26	.3240138	3.086288
.28	.3422837	2.921553
.30	.3605536	2.773513
.32	.3788236	2.639751
.34	.3970934	2.518299
.36	.4153633	2.407531
.3800001	.4336333	2.306096
.4000001	.4519032	2.212863
.4200001	.4701731	2.126876
.4400001	.488443	2.047322
.4600001	.5067129	1.973504
.4800001	.5249828	1.904824
.5000001	.5432527	1.840764

C = 1.6

P	× ipm	× mpi
.24	.3057439	3.270711
.26	.3240138	3.086288
.28	.3422837	2.921553
.30	.3605536	2.773513
.32	.3788236	2.639751
.34	.3970934	2.518299
.36	.4153633	2.407531
.3800001	.4336333	2.306096
.4000001	.4519032	2.212863
.4200001	.4701731	2.126876
.4400001	.488443	2.047322
.4600001	.5067129	1.973504
.4800001	.5249828	1.904824
.5000001	.5432527	1.840764

Table D.1   Continued

C = 1.8		
P	× ipm	× mpi
.24	.3057439	3.270711
.26	.3240138	3.086288
.28	.3422837	2.921553
.30	.3605536	2.773513
.32	.3788236	2.639751
.34	.3970934	2.518299
.36	.4153633	2.407531
.3800001	.4336333	2.306096
.4000001	.4519032	2.212863
.4200001	.4701731	2.126876
.4400001	.488443	2.047322
.4600001	.5067129	1.973504
.4800001	.5249828	1.904824
.5000001	.5432527	1.840764

C = 2		
P	× ipm	× mpi
.24	.3057439	3.270711
.26	.3240138	3.086288
.28	.3422837	2.921553
.30	.3605536	2.773513
.32	.3788236	2.639751
.34	.3970934	2.518299
.36	.4153633	2.407531
.3800001	.4336333	2.306096
.4000001	.4519032	2.212863
.4200001	.4701731	2.126876
.4400001	.488443	2.047322
.4600001	.5067129	1.973504
.4800001	.5249828	1.904824
.5000001	.5432527	1.840764

C = 2.2		
P	× ipm	× mpi
.24	.3057439	3.270711
.26	.3240138	3.086288
.28	.3422837	2.921553
.30	.3605536	2.773513
.32	.3788236	2.639751
.34	.3970934	2.518299
.36	.4153633	2.407531
.3800001	.4336333	2.306096
.4000001	.4519032	2.212863
.4200001	.4701731	2.126876
.4400001	.488443	2.047322
.4600001	.5067129	1.973504
.4800001	.5249828	1.904824
.5000001	.5432527	1.840764

Table D.1 Continued

C = 2.4		
P	× ipm	× mpi
.24	.3057439	3.270711
.26	.3240138	3.086288
.28	.3422837	2.921553
.30	.3605536	2.773513
.32	.3788236	2.639751
.34	.3970934	2.518299
.36	.4153633	2.407531
.3800001	.4336333	2.306096
.4000001	.4519032	2.212863
.4200001	.4701731	2.126876
.4400001	.488443	2.047322
.4600001	.5067129	1.973504
.4800001	.5249828	1.904824
.5000001	.5432527	1.840764

C = 2.6		
P	× ipm	× mpi
.24	.3057439	3.270711
.26	.3240138	3.086288
.28	.3422837	2.921553
.30	.3605536	2.773513
.32	.3788236	2.639751
.34	.3970934	2.518299
.36	.4153633	2.407531
.3800001	.4336333	2.306096
.4000001	.4519032	2.212863
.4200001	.4701731	2.126876
.4400001	.488443	2.047322
.4600001	.5067129	1.973504
.4800001	.5249828	1.904824
.5000001	.5432527	1.840764

C = 2.8		
P	× ipm	× mpi
.24	.3057439	3.270711
.26	.3240138	3.086288
.28	.3422837	2.921553
.30	.3605536	2.773513
.32	.3788236	2.639751
.34	.3970934	2.518299
.36	.4153633	2.407531
.3800001	.4336333	2.306096
.4000001	.4519032	2.212863
.4200001	.4701731	2.126876
.4400001	.488443	2.047322
.4600001	.5067129	1.973504
.4800001	.5249828	1.904824
.5000001	.5432527	1.840764

Table D.1   Continued

C = 3		
P	× ipm	× mpi
.24	.3057439	3.270711
.26	.3240138	3.086288
.28	.3422837	2.921553
.30	.3605536	2.773513
.32	.3788236	2.639751
.34	.3970934	2.518299
.36	.4153633	2.407531
.3800001	.4336333	2.306096
.4000001	.4519032	2.212863
.4200001	.4701731	2.126876
.4400001	.488443	2.047322
.4600001	.5067129	1.973504
.4800001	.5249828	1.904824
.5000001	.5432527	1.840764
C = 3.2		
P	× ipm	× mpi
.24	.3057439	3.270711
.26	.3240138	3.086288
.28	.3422837	2.921553
.30	.3605536	2.773513
.32	.3788236	2.639751
.34	.3970934	2.518299
.36	.4153633	2.407531
.3800001	.4336333	2.306096
.4000001	.4519032	2.212863
.4200001	.4701731	2.126876
.4400001	.488443	2.047322
.4600001	.5067129	1.973504
.4800001	.5249828	1.904824
.5000001	.5432527	1.840764
C = 3.4		
P	× ipm	× mpi
.24	.3057439	3.270711
.26	.3240138	3.086288
.28	.3422837	.2.921553
.30	.3605536	2.773513
.32	.3788236	2.639751
.34	.3970934	2.518299
.36	.4153633	2.407531
.3800001	.4336333	2.306096
.4000001	.4519032	2.212863
.4200001	.4701731	2.126876
.4400001	.488443	2.047322
.4600001	.5067129	1.973504
.4800001	.5249828	1.904824
.5000001	.5432527	1.840764

Table D.1   Continued

C = 3.6

P	× ipm	× mpi
.24	.3057439	3.270711
.26	.3240138	3.086288
.28	.3422837	2.921553
.30	.3605536	2.773513
.32	.3788236	2.639751
.34	.3970934	2.518299
.36	.4153633	2.407531
.3800001	.4336333	2.306096
.4000001	.4519032	2.212863
.4200001	.4701731	2.126876
.4400001	.488443	2.047322
.4600001	.5067129	1.973504
.4800001	.5249828	1.904824
.5000001	.5432527	1.840764

C = 3.8

P	× ipm	× mpi
.24	.3057439	3.270711
.26	.3240138	3.086288
.28	.3422837	2.921553
.30	.3605536	2.773513
.32	.3788236	2.639751
.34	.3970934	2.518299
.36	.4153633	2.407531
.3800001	.4336333	2.306096
.4000001	.4519032	2.212863
.4200001	.4701731	2.126876
.4400001	.488443	2.047322
.4600001	.5067129	1.973504
.4800001	.5249828	1.904824
.5000001	.5432527	1.840764

C = 4

P	× ipm	× mpi
.24	.3057439	3.270711
.26	.3240138	3.086288
.28	.3422837	2.921553
.30	.3605536	2.773513
.32	.3788236	2.639751
.34	.3970934	2.518299
.36	.4153633	2.407531
.3800001	.4336333	2.306096
.4000001	.4519032	2.212863
.4200001	.4701731	2.126876
.4400001	.488443	2.047322
.4600001	.5067129	1.973504
.4800001	.5249828	1.904824
.5000001	.5432527	1.840764

6.4.3 EFFECT OF PRECISION ON ACCURACY

In a 60-mpi soil, the water level falls half an inch during the 30-min measurement interval. So, with 1/16-in. precision, a test in a 60-mpi soil could yield measurements of:

    30 ÷ 7/16 = 68.5 mpi, or
    30 ÷ 8/16 = 60 mpi, or
    30 ÷ 9/16 = 53.5 mpi.

Note that their average is 61 mpi, instead of 60. If the average were obtained from ipm values, the reciprocal of this average would be exactly 60 mpi.

Also note that if the precision were to the nearest 1/8 in., the figures above would be 80, 60, and 48 mpi respectively, and their incorrect mpi average would be 63 mpi. The errors committed when averaging mpi directly instead of through reciprocals of ipm are rather small.

If three tests are made in a 60-mpi soil, and the precision is 1/8 in., just by chance all three tests might measure 48 mpi, or all might measure 80 mpi, or two might measure 60 and one 48 mpi, etc. So, we want to find out what proportion of average mpi values will overestimate or underestimate the actual mpi.

Before we can do that, we must find out what kind of statistics the measurements obey. Perhaps the clue can be found in the type of data I have encountered while reviewing perk reports, similar to the data shown below.

Below, the rates of fall of water level in three test holes are "stabilized" and show no trend. The last four sequential measurements (at 30-min intervals) are, in inches:

    Hole 1 . . . . .4/8, 4/8, 3/8, 4/8
    Hole 2 . . . . .4/8, 5/8, 4/8, 5/8
    Hole 3 . . . . .3/8, 4/8, 3/8, 4/8

Yes, most perk test measurements I have seen have a precision of 1/8 in. It is difficult to be more precise than that under field conditions (readings are affected by light reflections, angle of sight, meniscus curvature, water ripples, receding water level). (A precision higher than 1/16 in. can be easily achieved by installing a float within the perk hole. This float moves a vertical wire or rod up and down a scale, which can be easily read [see Figure 6.1], but only one of the local perk consultants has bothered to measure to 1/16 in.)

Well, the way I see the measurements above, it seems to me that each final reading has the same probability of being "on the mark" as of being 1/8 in. higher, or 1/8 in. lower. Assuming that this is the case, we can construct a table of chance deviations from a mean. These deviations are designated "+" or "−." The mean is designated "o." For example, the measurement from one hole might be +, or o, or -. The measurements from two holes might be tabulated as

	+	o	−
+	+ +	+ o	+ −
o	o +	o o	o −
−	− +	− o	− −

So we have that in one out of nine possible chance results obtained with two test holes, both holes are going to err on the + side. And in two out of nine, one hole will err on the + while the other will not err. And in two out of nine, one hole will err on the + side while the other will err on the − side, so errors will cancel out and yield a result without error, as if it were a oo situation. Adding up all the net + and the net oo for one, two, or more simultaneous tests (the net minuses are symmetric to the net pluses; there are as many − as +, as many −− as + +, etc.), we can easily construct Table D.2.

If the mpi of only one hole is measured, this mpi has one chance in three of being +, or −, or o. If two holes, 1/9 will have a net + +, 2/9 will have a net +, 3/9 will be oo (a + and a − average out to an oo), and so on.

The chance that the errors in each of the three holes are all on the + side is 1/27, or less than 4%. The chance that two out of three err on the + side is 3/27, or about 10%. The

Table D.2 Chance Deviations From a Mean[a]

	1	2	3	4	5
+ + + + +					1
+ + + +				1	5
+ + +			1	4	15
+ +		1	3	10	30
+	1	2	6	16	45
o	1	3	7	19	51
–	1	2	6	16	45
– –		1	3	10	30
– – –			1	4	15
– – – –				1	5
– – – – –					1
	3	9	27	81	243

[a]This triangle of numbers can be easily expanded. Please note that each number in the triangle can be derived by adding the number to its left to the number above and below this left number.

chance that one out of three errs on the + side is 6/27, or 22% (note that, in this case, since there are three holes, the average will be affected only one-third as much as if each had a + error). So if we designate a + or – error as "f", the average of the errors will be as shown below:

magnitude	probability
fff ÷ 3 = f	1/27
ff0 ÷ 3 = 2/3 f	3/27
f00 ÷ 3 = 1/3 f	6/27

Repeating the procedure for one through six holes, we can obtain the results shown in Table D.3.

Looking at the lower left portion of this table, we see that the measurement from one hole has a 0.33 probability of being off by the full 1/8 (or 1/16 in., whichever precision is being used). Let's use 1/8 in. The average from two holes has a 0.11 or 11% probability of being off by 1/8 in., and a 0.22 or 22% probability of being off by half of 1/8 in. In other words, of the averages of two holes in a 60-mpi soil, 11% will yield values as high as 80 mpi, 11% will yield values as low as 48

Table D.3   Magnitude of Error (and Its Probability of Occurrence, in Parenthesis)

F(1/3)	F(1/9)	F(1/27)	F(1/81)	F(1/243)	F(1/729)
	$\frac{1}{2}$F(2/9)	$\frac{2}{3}$F(3/27)	$\frac{3}{4}$F(4/81)	$\frac{4}{5}$F(5/243)	$\frac{5}{6}$F(6/729)
		$\frac{1}{3}$F(6/27)	$\frac{2}{4}$F(10/81)	$\frac{3}{5}$F(15/243)	$\frac{4}{6}$F(21/729)
			$\frac{1}{4}$F(16/81)	$\frac{2}{5}$F(30/243)	$\frac{3}{6}$F(50/729)
				$\frac{1}{5}$F(45/243)	$\frac{2}{6}$F(90/729)
					$\frac{1}{6}$F(141/729)
F(.33)	F(.11)	F(.037)	F(.012)	F(.0041)	F(.0014)
	$\frac{1}{2}$F(.22)	$\frac{2}{3}$F(.11)	$\frac{3}{4}$F(.049)	$\frac{4}{5}$F(.020)	$\frac{5}{6}$F(.0082)
		$\frac{1}{3}$F(.22)	$\frac{2}{4}$F(.123)	$\frac{3}{5}$F(.062)	$\frac{4}{6}$F(.029)
			$\frac{1}{4}$F(.198)	$\frac{2}{5}$F(.123)	$\frac{3}{6}$F(.069)
				$\frac{1}{5}$F(.185)	$\frac{2}{6}$F(.123)
					$\frac{1}{6}$F(.193)

mpi, 22% will yield values of 68.5 mpi, and 22% values of 53.5 mpi. When we get to the average of four holes, only 1.2% of these averages will be affected by the full error of 1/8 in., and 12.3% by 2/4 of 1/8 in. That is, in a 60-mpi soil, only 1.2% of the averages will yield values as high (low) as 80 (48) mpi, and 12.3% as high (low) as 68.5 (53.5) mpi.

The average of a minimum of four holes seems to be a reasonable standard, when the precision is 1/8 inch and the soil is perfectly uniform.

Winneberger (see Reference 3 in Chapter 6) requested five people to take five measurements each under simulated perk test conditions with a static water level. The average measurement error was 3/16 in. One of the five people, a very skilled technician, was more accurate than the rest.

6.4.4 THE "CAN" OR SOIL CYLINDER

Since the flow is proportional to pressure, the total flow or the decrease in h is proportional to the flow through the sidewall plus the flow through the bottom. The pressure on the sidewall is always 1/2 h, and the pressure on the bottom is h. Hence,

$$-dh/dt = ch + c(1/2\ h)$$
$$= 1.5\ ch$$

Now 1.5 c is a constant times a constant, which is another constant, say c″. Therefore, we get the same derivation as in page 79, only with a c″ instead of a c. And if we solve the equation with the same data points as to times and h levels, we get identical results.

(It is worthy of note that, if we conduct a real experiment, when h goes down near the bottom, the water will not flow exactly as the equation predicts because of the cohesion between water molecules and attraction between container and water molecules.)

6.4.5 EFFECT OF CHANGING HEAD

a. The erroneous statement is "The hydraulic head was increased 1.33 times." Only h, surface head, was increased this amount.

b. They were not proportional because when one starts with a small h, one develops a small X in a given time. So by increasing h suddenly, one develops a relatively large gradient, especially up near the sidewalls. These phenomena are rather predictable.

c. If h is much larger than the diameter: A sudden decrease in h should have no effect for a little while, but then it should stabilize to a rate proportional to the decrease. Conversely, in the short run, an increase should increase ipm more than proportionally, but then the ipm should tend to stabilize at a rate proportional to the increase in h. Chapter 9 makes this clearer.

6.4.6 PLUGGED BOTTOM

There are at least three possibilities. One, the bottom of the controls might have been compacted, so no difference was observed. Two, there might have been a thin, low-permeability clay or silt layer just below the holes, so that all the flows were mainly through the sidewalls. Probably neither of these instances occurred, because Weibel et al. knew about compaction risks and impermeable soil layers, and they would have taken these factors into account. Here is the third, most likely possibility:

The bottom area of a hole 4 in. in diameter is 12.5 in.[2] If we increase the radius only 1 in., the bottom area increases to 28 in.[2] The water had to move only a little bit sideways around the plug, and increase its path (or X) just a bit, to flow through an equivalent cross-sectional area, at a rate similar to that in an unplugged hole.

6.4.8 APPLICABILITY OF OLIVIERI-ROCHE (O-R) CORRECTIONS

The O-R hypothesis neglects the effect of differential flows per unit area of sidewall and of bottom, due to the differences in gradients. The surface head acting on the bottom is h, and on the sidewall, 1/2 h. Also, it assumes that the soils are homogeneous and isotropic, the same in all directions; often they are not, and the ratio of downward to sideways flows can vary greatly.

However, if the soils are fairly homogeneous and isotropic, and the conditions are such that the magnitude of h or 1/2 h does not affect the gradient, and h is kept constant, the O-R hypothesis should work for the common hole diameters. And so should the simple ratios of diameters, which yield almost the same results as the more elaborate O-R correction factors, numerically speaking.

I requested perk testers to test holes of different sizes, and I have also conducted tests of my own. In a very sandy soil, I tested a hole, carefully enlarged its diameter, and tested it again. The rates were within 9% of those predicted by O-R. I repeated the experiment in a dense clayey silt soil of very low permeability (about 200 mpi), and the rates were 30% off those predicted by O-R.

Van Kirk et al. (Chapter 6, note 8) analyzed perk data obtained by Winneberger in two different soils and found that O-R predictions were less than 10% off. On the other hand, either the data from local perk testers showed no differences due to hole diameter, or else the differences were roughly 60% as large as expected on the basis of O-R or of diameter ratios.

The clue to this puzzle might be found in an article recently published by Fritton et al.* They concluded that a theoretical equation

$$\log k = -\log mpi - \log (1 + 4/\pi ar)$$

could be used to convert mpi data to k values. To account for the experimental data, "a" varied from 0.1 to 100 per meter; "a" is a constant (specific for each soil) that may reflect soil anisotropy (nonuniformity, stratification). And r is the radius of the perk hole.

Well, I manipulated this equation and derived from it the formula below:

$$mpi''/mpi' = ([r + 4/\pi a]/r) \times ([2r]/[2r + 4/\pi a])$$

in which mpi' is the mpi measured in a hole of a given radius r, and mpi" is the mpi measured if the radius of this same hole were doubled.

According to this formula, if a 3-in. radius is doubled, the most that can happen is a doubling of the mpi (if a > 100); the least that can happen is no effect (the measured mpi stays the same after doubling the radius if a < 10). So, depending on the type of soil around the hole, a correction factor could be equal to 1, 2, or any value in between.

As far as I am concerned, the best policy to follow is:

---

*Fritton D.D., et al. 1986. Determination of saturated hydraulic conductivity from soil percolation test results. *Soil Sci. Soc. Am. J.* 50: 273–276.

a. Use a standard hole diameter; do not count on O-R or ratio of diameters for routine corrections of mpi's obtained in holes of various diameters.

b. If a hole must be much larger than standard, then correct in the most conservative direction. (For instance, a larger-diameter hole in a clay soil will yield slower absorption rates, so do not correct for diameter. But a larger-diameter hole in an excessively permeable soil will yield rates purporting to show the soil not to be excessively permeable; use ratio-of-diameter corrections in this case.)

Appendix E

9.5.4 SIZING A PIT

This is not a technical problem. It is a "public relations" problem.

My views are reflected in Chapter 20, and in the old saying, "When in Rome, do as the Romans do." . . . But if I had to do as the local sanitarians do, I'd write a polite note explaining an "alternative" (i.e., more reasonable) way of sizing a pit, for their consideration.

Appendix F

10.1 EXPERIMENTAL SYSTEM

The document recorded to enable experimentation is reproduced in Figure F.1.

N O T I C E

Regarding matter affecting property described as:

_____   _____
Assessor Parcel No.   *Legal Description*

_____

Address

in the vicinity of _____ , San Bernardino County.
                          Community

Above premises are encumbered by the following restrictions:

1. The premises are to be served by an experimental sewage disposal system
   until connected to sewers.

2. Should this experimental system be found not to perform in a satisfactory
   manner by the County Department of Environmental Health Services, operators
   and/or owners of such system and/or owners of property served by such
   system will discontinue its use and operation immediately and will install
   and operate a holding tank if legally and practically feasible or vacate
   the property.

3. Access shall be provided so that Environmental Health Services
   personnel may inspect any outdoor portion of experimental system
   at anytime. Observation pipes shall be installed as directed by
   Environmental Health Services and left unmolested.

4. The operators, and/or owners of property above described will hold the
   County of San Bernardino and any and all of its officers or representatives harmless and indemnify and defend for any of the consequences or
   liabilities that may arise from the County approving an experimental
   sewage disposal system (County Reference No. _____) in the
   above-described property. (This includes any costs resulting from
   enforcement activities.)

5. The owner(s) of the above-described property and his, her, or their
   successors or assigns are encumbered by all the restrictions herein
   specified.

   Agreed By: _____

   _____
   Signature(s) of Property Owner(s)

*(Notarize and record with County Recorder and submit a copy to the
Department of Environmental Health Services, Wastewater Management Section)*

Figure F.1   Public notice to be filed by persons using experimental sewage disposal systems, San Bernardino County, California.

Appendix G

EXCERPTS FROM THE CALIFORNIA ADMINISTRATIVE CODE, TITLE 22

66696. Toxicity Criteria.

(a) A waste, or a material, is toxic and hazardous if it:

(1) Has an acute oral LD50 less than 5,000 milligrams per kilogram; or

(2) Has an acute dermal LD50 less than 4,300 milligrams per kilogram; or

(3) Has an acute inhalation LC50 less than 10,000 parts per million as a gas or vapor; or

(4) Has an acute aquatic 96-hour LC50 less than 500 milligrams per liter when measured in soft water (total hardness 40 to 48 milligrams per liter of calcium carbonate) with fathead minnows (*Pimephales promelas*), rainbow trout. . . .

. . . .

66702. Ignitability Criteria.

(a) A waste, or a material, is ignitable and hazardous if it:

(1) Is a liquid, other than an aqueous solution containing less than 24 percent alcohol by volume, and has a flash point less than 60 degrees centigrade (140 degrees Fahrenheit), as determined by a Pensky-Martens Closed Cup Tester, using the test method specified in American Society for Testing and Materials (ASTM) Standard D-93–79, or a Setaflash

Closed Cup Tester, using the test method specified in ASTM Standard D-3278–73; or

(2) Is not a liquid and is capable, under standard temperature and pressure, of causing fire through friction, absorption of moisture or spontaneous chemical changes and, when ignited, burns so vigorously and persistently that it creates a hazard; or

(3) Is a flammable compressed gas as defined in 49 CFR 173.300(b) (codified October 1, 1982) and as determined by the test methods described in that regulation; or

(4) Is an oxidizer as defined in CFR 173.151 (codified October 1, 1982).

. . . .

66705. Reactivity Criteria.

(a) A waste, or a material, is reactive and hazardous if it:

(1) Is normally unstable and readily undergoes violent change without detonating; or

(2) Reacts violently with water; or

(3) Forms potentially explosive mixtures with water; or

(4) Generates toxic gases, vapors or fumes, when mixed with water, in a quantity sufficient to present a danger to human health or the environment; or

(5) Is a cyanide or sulfide bearing waste which, when exposed to pH conditions between 2 and 12.5, generates toxic gases, vapors or fumes in a quantity sufficient to present a danger to human health or the environment; or

(6) Is capable of detonation or explosive reaction if it is subjected to a strong initiating source or if heated under confinement; or

(7) Is readily capable of detonation or explosive decomposition or reaction at standard temperature and pressure; or

(8) Is a forbidden explosive as defined in 49 CFR 173.51 (codified October 1, 1982).

. . . .

66708. Corrosivity Criteria.

(a) A waste, or a material, is corrosive and hazardous if it:

(1) Is aqueous and has a pH less than or equal to 2 or greater than or equal to 12.5, or its mixture with an equivalent weight of water produces a solution having a pH less than or equal to 2 or greater than or equal to 12.5. The pH shall be determined by a pH meter using either test method 9040 specified in "Test Methods for Evaluating Solid Waste, Physical/Chemical Methods", SW-846, U.S. Environmental Protection Agency, 2nd edition, 1982, or as described in "Methods for Chemical Analysis of Waste and Wastes", EPA 600/4–79–020, March 1979; or

(2) Is a liquid, or when mixed with an equivalent weight of water produces a liquid, and corrodes steel (SAE 1020) at a rate greater than 6.35 millimeters. . . .

Appendix H

33.057 Soil Testing Requirements.
When required by the Director, soil percolation testing shall be done in compliance with the current Percolation Test Report Requirements adopted by the Department. The Director may establish other means for determining liquid waste application rates and charge such fees as are appropriate and authorized by this Code.

33.058 Soils Testing Administration.
Persons performing soils percolation tests for review by the Department shall be subject to the following requirements:
  (a) Testers shall be qualified as one or more of the following:
    (1) State of California Registered Civil Engineer.
    (2) State of California Certified Engineering Geologist.
    (3) Business firm employing or comprised of one (1) or more State of California Registered Civil Engineers or State of California Certified Engineering Geologists.
    (4) State of California Registered Sanitarian.
  (b) Demonstration to the Director of competence in soil percolation testing and local procedures.

243

33.059 Revocation by Department of Testing Privileges.

Any tester may have his testing privileges revoked or suspended for any one or more of the following causes: (1) if found by the Department to have lost the status which qualifies such person to perform percolation testing; or (2) if found to have falsified information submitted to the Department in a report(s) or correspondence; or (3) if found to have provided any other false information to the Department on a material question; or (4) if found generally to have performed in other than a diligent manner regarding any testing performed or reports filed with reference to this Code.

(a) Upon determining probable cause for revocation or suspension of testing privileges, the Department shall give written notice to the tester to show cause why his testing privileges should not be revoked or suspended.

(b) Upon written notification to show why his privileges should not be revoked or suspended, the tester may appeal to the Director within ten (10) working days for a hearing.

(c) Within five (5) working days after the close of any hearing, the Director shall notify the tester whether his privileges have been revoked or suspended. If the decision of the Director is to revoke or suspend the privileges, the notice or revocation or suspension shall state the grounds therefor.

(d) Unless special approval is granted by the Director, a person whose privileges have been revoked pursuant to this Section 33.059 may not reapply for reinstatement unless revocation was based solely upon loss of status which qualified the person for testing privileges and such loss of status was without wrongdoing on the part of such person.

31.061 Applicability of This Hearing Procedure.

Notwithstanding any other provision of this County Code, the following administrative hearing procedure shall be applied for any hearing pertaining to the suspension, revocation or reissuance of any license, permit certificate or entitlement where such action is provided for in any of the provisions of Title 3 of this Code except when a hearing before the Board of Supervisors is otherwise provided. This procedure provides an appeal from a permit suspension. Said appeal shall be made in writing to the Director within ten (10) working days of the permit suspension, denial or revocation and shall contain the address to which the notice of hearing shall be sent.

31.062 Hearing Officer.

The hearing officer for hearing pursuant to this chapter shall be the Director of the Department of Environmental Health Services or his appointee for such purpose. Any such appointee shall be a person who has no knowledge of the facts of the particular case at the outset of the hearing, and a person not immediately involved with regulation of the particular code provisions concerned.

31.063 Notice.

At least ten (10) days written notice of the hearing shall be given to the holder of the right prior to the hearing date. The hearing date may be postponed or continued by stipulation of the parties. If the party notified does not respond or appear, no further hearing procedure shall be required.

31.064 Hearing Procedures.

Witness shall swear or affirm to tell the truth. The oath or affirmation shall be taken by the hearing officer.

The enforcing officer shall present his case first, with oral testimony and documentary evidence or other exhibits. The responding party shall have the right to be represented by counsel, and shall have the right of cross-examination.

The responding party may present its response after the enforcing officer has presented his case. The enforcing officer shall have the right of cross-examination.

After both sides have completed presenting evidence, the enforcing officer may comment on the evidence and argue.

After the enforcing officer has commented on the evidence and/or argued, the responding party may do the same.

31.065 The Hearing Officer's Determination.

No determination or order shall be based solely on the basis of hearsay evidence.

The Hearing Officer shall make his determination within five (5) working days of the end of the hearing, unless the responding party stipulates to a greater period of time. The determination shall be in writing, and shall state the findings upon which the determination is made. Final determination is the responsibility of the Director, and shall be made in writing within five (5) working days of the Hearing Officer's report. There shall be no further non-court proceedings or appeal, unless specifically so provided elsewhere in this Code.

Appendix I

18.1 SOIL PERK REPORT STANDARDS

NOTICE: Before conducting perk tests, you must contact the County Specialist and provide the assessor's parcel number of the site to be tested, and the date of testing. At his option the Specialist may wish to conduct a field inspection during testing or shortly thereafter. Leave three-foot laths marked with your initials and hole number at each backfilled hole. The date when the Specialist (or his secretary) was contacted must be stated in the report.

[In the "old times," a perk consultant would test without notifying anyone. If he found the site unsuitable, he kept this finding confidential in order not to harm his client. After I added the notification requirement, I was able to find out what happened in the field, and why a perk test report was not forthcoming months after testing. Now that the notice above is in effect, perk consultants are very cooperative and reveal the existence of adverse site or soil conditions. Also, when a new perk consultant comes on board, I can be present to guide him during his first few tests. Thereafter, I drop by unannounced.]

Introduction

I. A perk report is required:
  a. On all subdivisions of land.

b. On any parcel or land division where current data will not allow the county specialist to set a sewage disposal rate.

c. On any single lot where space or soil conditions are critical.

d. For all septic systems within areas specifically defined by the governing authority. [Refer to an appendix, specific for every county.]

II. Those who prepare the perk report also assume responsibility for it. Preparers must have demonstrated knowledge and understanding of local criteria, requirements, and procedures for perk testing and perk report preparation and also must qualify under state law or county ordinance.

Reports must show the original signature and registration number of the preparer. Photocopies are not acceptable.

[There used to be a firm which allegedly used photocopied signatures of former employees in reports with falsified data.]

Format and Other Requirements

State when Specialist was given notice of proposed date of testing.

1. DESCRIPTION OF SITE AND OF PROPOSED PROJECT

1.1 PREPARED FOR: Name of client, address, phone.

1.2 LOCATION OF SITE: Assessor's parcel number, legal description, method of location in the field (client's word is not acceptable). Provide a clear sketch showing "how to get there"; give street addresses nearby, if any; point out landmarks if difficult to find. It is the report preparer's responsibility to ensure tests are conducted where described in the report.

1.3 PROPOSED DEVELOPMENT

a. Type of project: condominium, apartments, subdivision tract, shopping center, etc.

b. Acreage, number of lots, average and range of lot sizes.

c. Type of sewage disposal: leachline, seepage pits, discharge to separate (single) or common (confluent) septic systems.

1.4 VERBAL DESCRIPTION OF PROPERTY

a. Slope/grading.

b. Floodway, flood plain, streams, and/or drainage courses.

c. Vegetation type and density (especially indicators such as willows, cattails, cactus, green patches).

d. Existing structures (including septic systems); general evaluation of surrounding area and density of development.

e. Rock outcroppings. Specify type of rock.

f. Groundwater table information. Specify source of information.

g. Any other feature that may affect sewage disposal: springs, fill, obvious signs of slope instability, spots of vegetation, fractured bedrock, root channels, or cracks in the soil profile.

h. If any portion of the soil absorption system or reserve will not be located on natural soil ($\pm$ 1 ft of cut/fill for leachlines, 3 ft for seepage pits), provide a grading plan or title perk report, "preliminary perk report."

1.5 GRAPHIC DESCRIPTION OF PROPERTY (PLOT PLAN)

Plot plan must be to scale and include:

a. Contours. Provide a topo map unless site and surroundings are flat or have a uniform, constant slope (for instance, uniform slope of 5% downward from north property line to south property line). See Table I.1.

b. Floodways, floodplains, streams or drainage courses.

c. Significant vegetation (including trees near proposed leachline area).

Table I.1 Required Contour Intervals for Plot Plan

% slope	Maximum interval of contours, ft
0–2	2
3–9	4
≥ 10	10

d. Existing wells or remnant of wells on or within 200 ft of the property. Large high-density projects may impact wells even farther away.

e. Existing structures.

f. Rock outcroppings.

g. All borings or excavations (including those of tests that failed). If the report recommends that the septic system be installed in the area tested rather than in the general area, test borings shall be accurately dimensioned to property lines.

h. The proposed sewage disposal system. If none is contemplated in lots zoned for single homes (lot-sales subdivision), an hypothetical system for a five-bedroom home shall be shown to fit in the smallest or most difficult lot; if zoned for multi-unit development, the hypothetical system shall be shown to suffice for the effluent discharged by an average of 3 bedrooms per unit.

1.6 GRADING Where grading is expected, include a grading plan. If grading plan was prepared by others, perk report preparer must comment on its adequacy.

2. EQUIPMENT

List equipment used in detail. Where the soil is stratified, and low-permeability layers like clay or caliche may affect the leachline system, the soil profile shall be described by looking at it directly, in a backhoe trench, road cut, or suitable large (≥ 1 ft diameter) boring.

3. METHODOLOGY AND PROCEDURES

3.1 *Borings and trenchings*

a. *Distribution*: If not randomly distributed (grid method), state specific reasons for choosing selected locations.

b. *Depth*: The minimum depth of exploratory borings/trenches shall be to leachline depth plus 8 ft, and to seepage pit depth plus 10 ft, or in accordance with stricter local water quality control board requirements.

c. *Number*: For a single lot with a single dwelling unit, in an area with slight site limitations, the minimum number of exploratory borings/trenchings shall be one; the minimum number of test holes shall be four for leachline, and two for seepage pit use. (But two-thirds of tests must yield satisfactory results; more than the minimum number will be required if soil conditions are not optimal.) For all other types of development, or if site is in a general area of moderate or severe limitations, check with county specialist. The minimum number of tests is locality-specific.

3.2 *Standard test for leachlines*

The USPHS-EPA methodology is adopted for lack of a better alternative. The county specialist may allow a nonstandard methodology in addition to this one.

Excavation:

Bottom of excavation is approximately 13 in. above the expected bottom of the leachline trench.

Perk test hole is drilled or augered at the bottom of the excavation.

Hole:

a. Final diameter is 5.5 to 6.5 in. after scraping. Larger sizes are acceptable, as they yield slower infiltration rates. (Up to 8 in., no corrections are necessary.) But in coarse soils with mpi lower than 8, apply ratio-of-diameter corrections to mpi if the diameter is larger than 8 in. [See Section 6.4.8 in Chapter 6, and Appendix D (6.4.8), for explanation of corrections.]

b. Depth is 13–14 in.

c. Place 2 in. of gravel over bottom. A perforated tin can may be placed over the gravel.

Soaking:

Fill hole with 12 in. of clear water (10 in. above the gravel or bottom of perforated can).

a. If twice 10 in. seeps away in less than 10 min and soil is coarse-textured, testing can be conducted immediately; otherwise,

b. Maintain level (8–16 in.) for at least 4 hours (or until 5 gal have been absorbed [invert a full 5-gal bottle over the 8–10-in. level after ensuring the bottle is well secured and surges will not scour sides of hole]).

[In coarse-textured soils, with mpi 1–4, the bottle will be empty after about 1 hr or less; in 15-mpi soils, after about 7 hr; in 120-mpi soils, the bottle will not be empty the following day.]

Testing:

Except as noted in (a) and (b) above, begin testing after 15 hr and finish within 30 hr after beginning of soaking. Refill after each measurement. Measure from a fixed reference point.

[If there is still at least 6 in. of water in the hole (4 in. above the gravel) after 5 gal have been absorbed, or after 15 hr from start of soaking, remove the bottle, restore water level to 8 in., and make at least 2 final measurements; the interval of the measurements is modified so that the decline in water level is kept within 1 to 3 in. Otherwise follow usual procedure:]

a. Fill hole to exactly 8 in. from bottom of hole (6 in. from top of gravel or bottom of can).

b. If 6 in. is gone in 30 min, use 10-min measurement intervals; otherwise, 30-min intervals.

c. Measure to nearest 1/16 inch. Lower precision may be acceptable if results justify such imprecision.

d. Make at least six consecutive measurements until three do not vary by more than 1/16 in. (Lower precision may be acceptable if justified.) The interval of the final three measurements is modified so that the decline in water level is kept between 1 and 3 in.

e. Where gravel packing or similar measures are taken to

prevent soil migration or sloughing off, some additional holes should be soaked and used as controls to see if such hazard can occur and if special measures are needed during construction of the leachfield.

3.3 Standard test for seepage pits

There is no standard test for seepage pits. See Chapter 9. What follows is the usual falling head test procedure, with an improved formula.

a. Drill to proposed depth of pit. Hole is 6 to 8 in. in diameter. (Exploratory borings may be backfilled 10 ft and used for testing provided top of fill is fine-textured or is sealed with driller's mud and protected with 1 ft of gravel.)

b. Fill hole with clear water to the level of inlet of assumed pit.

c. In highly sandy soils, where water on two consecutive fillings seeps faster than half the wetted depth in 30 min, measurement intervals shall be 10 min and the test shall be run for at least 1 hr until three consecutive readings do not vary by more than 10%. Refill after each measurement interval. Decomposed granite is not to be considered a sandy soil.

d. In other soils, soak the hole and let set overnight. The following day fill to assumed depth of pit inlet. From a fixed reference point, measure drop in water level over 30-min intervals for at least 5 hr until readings do not vary by more than 10%. Refill after each reading but the last. After each reading, measure the depth to bottom.

Caving in excess of 15% of depth may invalidate results. Use gravel packing in at least one test hole and at least 20% of test holes.

4. RESULTS

4.1 Soil logs

A boring or trench log shall be submitted for each exploratory boring or trench.

a. Texture. If you use the U.S.C. classification, state approximate percentage of sand, silt, and clay. Otherwise, use the classification in Table I.2, based on handling and appearance.

b. Colors. Specify if dry or moist soil. Note reduction-oxidation mottling.

c. Presence and extent of small/large roots.

d. Ease of excavating/drilling (soft, firm, hard, refusal).

e. Moisture. If moisture is found, allow 24 hr to determine if free water will appear.

f. Free water.

g. Other.

4.2 Test results

4.2.1 Tabulate all final results, including those of tests that "failed" to meet standards.

4.2.2 Provide copies of all field data and calculations, using the following format:

a. *Leachline test.* Include hole number, depth of bottom below grade, type of strata tested, method to prevent side-wall caving, hole diameter, hours soaking, name of tester, date tested, condition of hole (caving or siltation?) See Figure I.1.

b. *Seepage pit, falling head test.* Include boring number, diameter of boring (feet), depth of bottom below grade, hours presaturation, name of tester, date tested, and strata peculiarities (if any). See Figure I.2.

5. DISCUSSION OF FINDINGS

5.1 Discuss uniformity or variability of results. A uniform soil unit is delineated by at least four test results falling within one-third of their mean mpi.

At any given location with presumably uniform soils, at least two-thirds of tests must show acceptable mpi (three out of four, or four out of six).

5.2 Discuss possible sources of error or variability of

Table I.2   Soil Textural Classes[a]

TEXTURAL PROPERTIES OF MINERAL SOILS

Soil Class	Feeling and Appearance	
	Dry Soil	Moist Soil
Sand	Loose, single grains which feel gritty. Squeezed in the hand, the soil mass falls apart when the pressure is released.	Squeezed in the hand, it forms a cast which crumbles when touched. Does not form a ribbon between thumb and forefinger.
Sandy Loam	Aggregates easily crushed; very faint velvety feeling initially but with continued rubbing the gritty feeling of sand soon dominates.	Forms a cast which bears careful handling without breaking. Does not form a ribbon between thumb and forefinger.
Loam	Aggregates are crushed under moderate pressure; clods can be quite firm. When pulverized, loam has velvety feel that becomes gritty with continued rubbing. Casts bear careful handling.	Cast can be handled quite freely without breaking. Very slight tendency to ribbon between thumb and forefinger. Rubbed surface is rough.
Silt Loam	Aggregates are firm but may be crushed under moderate pressure. Clods are firm to hard. Smooth, flour-like feel dominates when soil is pulverized.	Cast can be freely handled without breaking. Slight tendency to ribbon between thumb and forefinger. Rubbed surface has a broken or rippled appearance.
Clay Loam	Very firm aggregates and hard clods that strongly resist crushing by hand. When pulverized, the soil takes on a somewhat gritty feeling due to the harshness of the very small aggregates which persist.	Cast can bear much handling without breaking. Pinched between the thumb and forefinger, it forms a ribbon whose surface tends to feel slightly gritty when dampened and rubbed. Soil is plastic, sticky and puddles easily.

Table I.2 Continued

	TEXTURAL PROPERTIES OF MINERAL SOILS	
Soil Class	Feeling and Appearance	
	Dry Soil	Moist Soil
Clay	Aggregates are hard; clods are extremely hard and strongly resist crushing by hand. When pulverized, it has a grit-like texture due to the harshness of numerous very small aggregates which persist.	Casts can bear considerable handling without breaking. Forms a flexible ribbon between thumb and forefinger and retains its plasticity when elongated. Rubbed surface has a very smooth, satin feeling. Sticky when wet and easily puddled.

[a]Source: EPA-625/1-80-12.

results. Siltation or caving of test holes may require special construction measures to prevent the soil absorption system from suffering the same fate.

6. DESIGN RECOMMENDATIONS

6.1 Criteria

a. For uniform soil units, use a mpi between mean and most-conservative. If there are no uniform soil units, the location with the least favorable mpi overrides any other test in the area of the disposal field.

b. Unless an area has been defined to have degraded or degradable groundwater, there shall be a minimum of 5 ft (leachlines) or 10 ft (seepage pits) of soil between the bottom of the soil absorption system and groundwater. If a soil has a perk time between 1 and 5 mpi, then the soil for a total thickness of 5 ft below the bottom of a leachline shall contain at least 10% of material passing the #200 U.S. Standard Sieve

t_1	depth$_1$	t_2	depth$_2$	Δt	Δd	$\Delta t/\Delta d$ mpi

Figure I.1 Form for leachline test.

d_b	t_i	t_f	t	d_i	d_F	$F = d_F - d_i$	$L = d_b - \dfrac{d_i + d_F}{2}$	$Q = 45FD/Lt$ or "pit mpi" $= 20FD/Lt$

Where d_b = depth to bottom, feet
  t_i = initial time, when refilling is completed, hour
  t_F = final time, when measurement is made
  t = time interval, 0.5 or 0.166 hour
  d_i = depth to water surface at t_i, feet
  d_F = depth to water surface at t_F, feet
  L = average length of water column, feet
  Q = gallons of water absorbed per square foot per day
  D = diameter of hole in feet

Figure I.2 Form for seepage pit, falling head test.

(and less than one-fourth of a representative vertical soil cross-section shall be occupied by stones larger than 0.5 ft). Where this requirement is not met, a 40-ft separation shall be maintained below the bottom of the leachline and highest expected groundwater (50-yr height). Fairly uniform coarse-textured soils shall not be used for seepage pits if the "pit mpi" is less than 10 and the separation to groundwater is less than 40 ft.

6.2 *Conversion of measurements to absorption area requirements.*
[Table I.3 was derived by interpolation of data from the EPA Manual Table 7–2. The requirements per 100 gal of septic tank capacity refer to septic tanks not larger than 1,500 gal; if larger tanks are used, the requirements should be increased in proportion to the ratios of rated flows to the tank capacities.]

For leachlines, use absorption areas in Table I.3 and leave reserve area for another set of leachlines. For pits 15 to 30 ft deep in unstratified or poorly stratified soils, use a sewage disposal rate between 1 gal/ft^2/day (if the soils are coarse-textured) and 0.4 gal/ft^2/day (if not coarse-textured but the falling head test Q is over 15 gal water/ft^2/day). If the soils are

Table I.3 Perk Times and Absorption Area Requirements

Perk time, mpi	Ft² of absorption area/100 gal of septic tank capacity
0–4	55
5–6	65
7–9	75
10–14	85
15–19	100
20–29	115
30–44	130
45–60	145
> 60	normally unsuitable; consult with county specialist[a]

[a]I prefer not to give a categorical "no" when I see mpi's higher than 60. Often, the measurement technique is at fault. Soils with mpi 60–120 and even higher can be used with increased absorption area. Also, when perk testers call in for consultations, they reveal where a problem area exists.

stratified and/or Q is less than 15 gal water/ft²/day, anything might happen regarding longevity. [A jurisdiction should not grant approval to seepage pits under these conditions, unless the perk report consultant assumes responsibility if the pit lasts less than 10 yr, or other arrangements are made.] Leave enough expansion area for at least three more sets of pits.

6.3 Special situations

a. If leachlines or pits serve a common system for two or more units, add 30% more square footage.

b. If leachlines must be under dirt driveways or pavement, increase affected footage by 30%. Such installations are *not* recommended. Ensure against subsidence and breakage. Provide aeration ports for leachlines.

c. If leachlines are affected by conditions (a) and (b) above, increase their footage by 60%.

d. For laundromats, and confluent systems serving mobile home parks or shopping centers (3 or more shops), multiply square footage by 2.5.

7 PLOT PER CURRENTLY ADOPTED UPC

Plot system and 100% (300%) expansion system. Slope precision is ± 2%. Remember UPC requirements are minimum requirements under the most ideal conditions. Disregard UPC Table I-4. Where higher requirements are called for, use higher requirements.

8 GENERAL DISCUSSION, CONCLUSIONS, AND RECOMMENDATIONS

8.1 Specify pertinent Water Quality Control Board requirements and state whether they are being met.

8.2 State unequivocally whether each lot has sufficient area to handle the liquid waste without creating a nuisance or contaminating groundwater. Include a qualifying statement if swimming pools, building expansions, etc., are or may be planned. Provide a set of septic system user guidelines to your client, whether prepared by yourself or by the county.

[End of Perk Standards text]

18.2 PRACTICAL CONSIDERATIONS

The previous section is useful as a framework. All possible constraints might not have been mentioned. The perk consultant must ensure that they are.

The main purpose of the perk report is not to force perk consultants to comply with a bunch of requirements, but to ensure that the septic system will "work" (i.e., will last at least 20 years, and will not cause health or pollution hazards). Other purposes are:

a. After reviewing the report, the county specialist may offer advice on how to improve a design.

b. A record is provided for possible action against unprincipled or incompetent consultants.

c. If something goes wrong and the septic system "fails"

prematurely, a consultant has something on record to prove that he did notice all physical constraints and that his recommendations were "state of the art."

When I evaluate a site, this is what I notice in reference to some of the subsections in the Perk Standards above:

1.4. *Description of property*

c. Vegetation is a very good indicator of rainfall, soil, and drainage conditions. Dwarf trees often indicate the presence of a shallow, low-permeability claypan; willows, high seasonal groundwater; cattails, year-round high groundwater.

d. Too many septic systems in an area might raise groundwater level or result in mounding.

e, f. Indicator-types of rocks a consultant should be familiar with are:

Shale, slate, and schist. They are associated with low-permeability/mounding and with fractures/channels. Soils derived from mica schist may yield erratic perk test readings. (Mica flakes are suspended when water scours sidewalls of test holes, and plug up the hole absorption surface.) Dolomite is associated with fractures/channels in humid climates. (Apparently, it is not a problem in California.)

1.5. *Determining slope*

If a hand level is not available, a carpenter's level and a tape can be used to determine slope. First pour some water on the soil and see where it runs, to ensure your line of sight will be parallel to the maximum slope. Tie the carpenter's level over the middle of a straight 2×4 stud. Place the stud level over the path of the water, with one end on the soil and the other up in the air. Measure the vertical distance from the end up in the air down to the soil. Divide this distance by the length of the stud and you get the slope.

If water is not available, move the floating end of the stud from side to side, and measure the longest vertical distance down to ground level.

3.2. *Instructions for technicians: leachline test*

Excavation: Bottom of excavated trench, approximately 13 in. above designed bottom of leachline trench.

Hole: Diameter after scraping, 5.5–6.5 in. Depth, 13 in.; gravel, 2 in. of pea gravel. Place perforated can on top.

Soaking: Fill hole with 12 in. of water (10 in. above bottom of perforated can).

a. If twice 10 in. seep away in less than 10 min and soil is coarse-textured, begin testing; otherwise,

b. Maintain level (8–13 in.) for 4 hr or 5 gal.

Testing: Begin testing 15–20 hr after beginning of soaking or after 5 gal are absorbed.

a. Fill hole to 8 in. level (6 in. from bottom of can)

b. If 6 in. are absorbed in 30 min, take readings at 10-min intervals; otherwise, at 30-min intervals.

c. Measure to nearest 1/16 in.

d. Make at least six consecutive readings until two measurements do not vary by more than 1/16 in.

e. Final three measurements: a decline of not less than 1 in. and not more than 3 in.; modify interval accordingly.

Simplified or shortcut procedure: Invert and secure full 5-gal bottle over hole so it discharges at the 8–10-in. level. After 5 gal are gone or the following day, remove bottle, and proceed with measurements. If there is still 6–8 in. of water in the hole, refill and make two final measurements; if there is less than 6 in. of water, go to "Testing," point (d), above.

5.1 *Uniform soil units*

A uniform soil unit is defined by a minimum of four (points or) measurements and a range of ± one-third of the mean. Measurements of 30, 40, 40, and 50 mpi define a uniform soil unit with a mean of 40 mpi.

Measurements of 20, 40, 60, and 80 mpi have a mean of 50 mpi. One-third of 50 is 16 mpi. Only two of these four measurements fall between 34 and 66 mpi. We do not have a uniform soil unit.

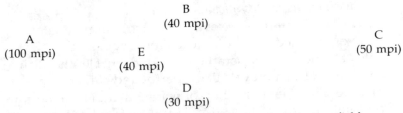

Figure I.3 Bird's-eye view of distribution of perk times in a field.

6.1 *Design criteria*

Let us assume mpi measurements are spread out in the field as shown in Figure I.3. The set of measurements A, B, C, D does not define a uniform soil unit. If E does not exist, the area ABCD is assigned 100 mpi. If E exists, the areas AEB and AED are assigned 100 mpi; the area BCDE is assigned an mpi between 40 and 50, say 45 mpi.

6.3 *Special situations*

To lower a groundwater table, or to divert springs away from leachline areas on sloping ground (> 5%), I have found no need for formulas or experimental hydraulic conductivity determinations. It is much more practical to install a simple French drain upslope (20 ft minimum horizontal distance upslope from leachfield; 25 ft minimum distance from each side of leachfield; and a bit lower than the maximum allowed depth of groundwater).

Appendix J

A leaflet prepared for public distribution is reproduced here. It was developed so that it could be printed and distributed by private businessmen (septic pumpers, real estate agents; see upper right corner of its first page) at no cost to the county. It was kept small and trim so that it would be read.

COUNTY OF SAN BERNARDINO
ENVIRONMENTAL
PUBLIC WORKS AGENCY

Revised 5/86

DEPARTMENT OF ENVIRONMENTAL
HEALTH SERVICES
385 North Arrowhead Ave
San Bernardino, CA 92415-0160

Distributed by courtesy of:

$ SAVE MONEY $

Mr./Ms. Homeowner:

This notification could save you anywhere from about $600.00 to the actual value of your home.

Unless you are already paying monthly sewer bills, you may be using a septic system for disposing of your wastewater (toilet flushings, shower, kitchen, washing machines). If and when a part of this system "fails", the health of the members of your household and of your neighbors is endangered; and then, you are required to replace the "failed" portion of your septic system.

This leaflet explains how to take good care of your septic system and avoid problems and costly replacements.

The septic system is composed of a septic tank and a leach field. The septic tank is usually a concrete "box" of about 1,000 gallons capacity. It is most often located about five feet from your house, under two or three feet of soil. Wastewater flows into this tank, and stays there temporarily. During this stay, the oil and grease in the wastewater rise to the top of the tank, where they form scum, and the solids sink to the bottom, where they form sludge. The clear wastewater in the middle of the tank flows to the leachfield, and percolates down into the soil.

The leachfield usually consists of either leachlines, which are gravel-filled underground trenches, or of seepage pits, which are vertical holes in the ground, four to six feet in diameter and fifteen to forty feet deep, with concrete block walls and a soil-covered lid on top.

If the scum and sludge are left to accumulate in the tank too long, they fill up the tank, are discharged into the leachfield, and plug up the soil. Then, the leachfield "fails": the waste-water comes up to the soil surface, and new leachlines or seepage pits must be constructed under a permit from the County (or City) Department of Building and Safety.

There are nine important principles that ensure a good and economic functioning of the septic system:

- 1 -

1. The most important one is to use the services of a county-licensed septic tank pumper every two to four years. The tank should be pumped when the total depth of floating scum plus bottom sludge exceeds one-third of the depth of the liquid in the tank. Check your phone book yellow pages for licensed Septic Tank Pumpers or request (free) referrals from the Liquid Waste Management Association -- *(714) 883-8701 (at 3972 N. Waterman Ave., Suite 106, San Bernardino 92404).*

   Make sure the second compartment of your septic tank is pumped at least every third time. (Uncover two lids.)

2. Although permitted by the Uniform Plumbing Code, it is better not to pave over. drive over, or trample your leachlines. Traffic vibration may also damage your seepage pits.

3. Do not waste money on yeasts, bacteria, or enzymes. According to scientific tests, they do not prevent premature failure of leachfields; only regular tank pumpings do.

4. Chemicals used to correct failures postpone final failure for only a few months. Often they corrode your septic tank and seepage pit lining and result in their collapse. Some are hazardous and contaminate groundwater. They are not recommended.

5. Do not flush down excessive amounts of oil or frying-pan grease. Do not leave faucets running for long periods of time. Keep faucets and toilets in good repair. If you use your garbage disposal your septic tank will fill up much faster.

6. Try not to destroy an old, failed leachfield: It may be used again after it rests for three to five years. When you have a new leachfield built, have a diversion valve installed to permit you to discharge wastewater to the new or to the old field. After three to five years, discharge to the new field on even-numbered years, and to the old one on odd-numbered years. If you let a leachfield rest every other year, the leachfield might last fifty or more years, with regular septic tank pumping. But if you have trees nearby, roots may plug up the drying leachfield.

7. If your contractor has not provided you with one, make a sketch of the layout of your septic system. (It will cost more to pump your tank or to install a new leachfield if the pumpers or contractors have to dig and search for the field or tank). Your licensed pumper can help you sketch the layout.

8. Keep a record of installations and of pumpings. If a leachfield fails less than five years after it has been installed, this may indicate that the leachfield was not constructed properly. Please report this to *Environmental Health Services, (714) 383-2543.*

9. When you sell (or buy) a house using a septic system, give (or request) the layout of septic system installations and record of pumpings as shown in this leaflet.

- 2 -

ADDRESS: _____

_____

		CONTRACTOR	
SEPTIC TANK			
Date of Installation	Size (gallons)	Name & Signature (initials)	Phone

LEACHLINES			
Dates of Installation	Length, Width, Depth of Rock		

SEEPAGE PIT(S)			
Dates of Installation	Depth		

DATE	LICENSED PUMPER NAME & PHONE
	Seepage Pit(s) Pumpings
	Septic Tank Pumpings

SKETCH OF SEPTIC SYSTEM LAYOUT

Appendix K

PROFESSIONAL INFORMATION SOURCES

Much information about septic systems can be found in other currently available publications. I believe that four of these should be in the library of every professional:

EPA's *Design Manual: On-site Wastewater Treatment and Disposal Systems*. Available from National Technical Information Service, 5285 Port Royal Road, Springfield, VA 22161; the order number is EPA-625/1–80–012. It is as essential to a septic systems consultant as a Bible is to a preacher.

J. T. Winneberger's *Septic Tank Systems*. This book points out what's wrong with the "Bible" above, and what's wrong with many common beliefs. It is also witty and humorous. It requires little math. You'll be delighted with Winneberger's writing style. (I was.) Order it from Butterworth Publishers, 80 Montvale Avenue, Stoneham, MA 02180; the order number is 0–250–40651–9, Septic Tank Systems/Winneberger.

R. Laak's second edition of *Wastewater Engineering Design for Unsewered Areas* is a good companion to Winneberger's book and vice versa. Laak's book is broad in scope (but very succinct), uses more math, and allows the reader to quickly find much information. It describes innovative infiltration and denitrification systems patented by its author. It was published in 1986 by Technomic Publishing Co., Inc., 851 New Holland Avenue, Lancaster, PA 17604.

IAPMO's Uniform Plumbing Code, *if* adopted by your local jurisdiction. IAPMO, 5032 Alhambra Avenue, Los Angeles, CA 90032.

At another level, the book *Ethics* is useful as a guide for consultants, and as a guide for governmental officials who want to know what to expect and demand from consultants. Although written for engineers, the book is pertinent to professionals in all environmental fields. (Gunn, A. S., and Vesilind, P. A. 1986. *Environmental Ethics for Engineers.* Lewis Publishers, 121 S. Main Street, Chelsea, MI 48118.)

Professionals who wish to keep abreast of practical developments in the field without spending too much time or money may wish to purchase the "Proceedings of the [nth] National Symposium on Individual and Small Community Sewage Systems," published every three years by the American Society of Agricultural Engineers. (The Proceedings of the Fourth National Symposium were published in 1985.) ASAE, 2950 Niles Road, St. Joseph, MI 49085-9659.

Professionals who wish to explore related theoretical topics, well beyond practical needs, might find the following papers useful as a starting point.

Corey, A. T., and Klute, A. 1985. Application of the potential concept to soil water equilibrium and transport. *Soil Sci. Soc. Am. J.* 49:3–11. (A review of fundamental concepts, with corrections and clarifications.)

Dane, J. H., et al. 1986. Estimating soil parameters and sample size by bootstrapping. *Soil Sci. Soc. Am. J.* 50:283–287. (A new statistical technique.)

Byers, E., and Stephens, D. 1983. Statistical and stochastic analysis of hydraulic conductivity and particle size in a fluvial sand. *Soil Sci. Soc. Am. J.* 47: 1072–1081. (Hydraulic conductivity was log-normally distributed, whereas the 10% finer mean particle sizes were normally distributed. Empirical formulas relating d[10] to k are useful only as rough approximations, not as predictors of k values.)

Boast, C. W., and Langebartel, R. G. 1984. Shape factors for seepage into pits. *Soil Sci. Soc. Am. J.* 48:10–15. (One way to calculate k in the field is to measure how fast the groundwater level recuperates after pumping a pit that reaches below the water table. This article explores the effect of pit shape and other factors.)

Russo, D., and Bresler, E. 1981. Soil hydraulic properties as a stochastic process. Part I. *Soil Sci. Soc. Am. J.* 45:682–687.

Russo, D., and Bresler, E. 1982. Soil hydraulic properties as a stochastic process. Part II. *Soil Sci. Soc. Am. J.* 46:20–26.

(At any location, values of k and of other soil properties are not independent of values obtained near that location. The number of samples may be reduced without decreasing accuracy.)

Practically all of the knowledge in the previously noted articles has been distilled into *Methods of Soil Analysis, Part I* (second edition, 1986). This book is indispensable to those who wish to understand soil testing methodology and its limitations. It discusses, in detail, just about every method to measure Darcy's constant k in the field and the laboratory. It is available from the Soil Science Society of America, 677 South Segoe Rd., Madison, WI 53771.

A recent book by Perkins (*Onsite Wastewater Disposal*, 1989, R. Perkins, Lewis Publishers, Inc.) has useful pictures and schematics of septic systems and components. I'd caution readers to ascertain the limitations of the easy-to-follow rules or recommendations provided in Perkins' book. For instance, my local experience is that the minimum amount of square feet of bottom area in Perkins' Table 5.1 regarding fine and coarse sand has commonly resulted in failures after about three years of use. As another instance, in the "Composting Toilets" section, Perkins describes the advantages of well-maintained composting toilets. Then he warns that a poorly maintained toilet will "stink and be an unpleasant mess to clean up." He states, "Before you install a compost toilet, recognize that it is a biological system just like your own body, and commit to the regular care and feeding of that system." Personally, I am more concerned about the survival of pathogens and parasites than about the smell and the unpleasantness. Moreover, I figured out that while it takes me one to five minutes to make use of a conventional toilet, it would take me an additional 10 to 20 or more minutes to collect and add the required sawdust or straw and to mix this into the compost after every use. I might exert myself for a day or two, but thereafter I would get tired of the maintenance hassle. I do not recommend allowing the installation of private composting toilets wherever the prospective users might feel as I do.

Constructed Wetlands for Wastewater Treatment (D. Hammer, ed., 1989, Lewis Publishers, Inc.) may be of interest to consultants who are tempted to treat sewage in natural wetlands or in artificial "single-home-size" mini-wetlands. Maintenance and vector problems prevent artificial mini-wetlands from becoming very common in Southern California. But in other parts of the United States, they might have a very bright future. The same can be said for artificial wetlands large enough to serve a whole community.

Appendix L

QUESTIONS, PROBLEMS, AND BRAIN TEASERS

1. The stabilized perk rate in a leachline perk test hole is 0.1 ipm. The measurement interval is 30 min. The soil is isotropic. If the measurement interval is shortened to 15 min, the measured perk rate usually will:

a. Double to 0.2 ipm.

b. Increase to a bit more than 0.1 ipm, but not as much as 0.2 ipm.

c. Stay the same.

d. Decrease to a bit less than 0.1 ipm, but not as much as 0.05 ipm.

e. Decrease to 0.05 ipm.

2. A technician has been sloppy and his readings of stabilized rates have a 1/4-in. precision. A trend is not detectable. If you had to use his findings, how would you make them more meaningful? His measurements for four test holes (six sequential measurements per hole) are given below. The inches of fall in 30-min intervals are:

Hole No.

1	2, 1.75, 2.25, 1.75, 1.75, 2.25
2	1, 1.5, 1.5, 1, 1, 1.5
3	2.25, 1.75, 2, 1.75, 2, 2.25
4	1.5, 1, 1, 1.5, 1, 1.5

3. A perforated pipe 2 in. in diameter and 34 ft long is

placed in a seepage-pit test hole 34 ft deep and 6 in. in diameter. Then the hole is gravel packed 30 ft high. The water level is refilled to 4 ft below grade (30 ft above bottom). The gravel voids are 33%. If the rate of fall is 5 ft during a measurement interval of 30 min,

a. What is the true rate of fall after accounting for the gravel?

b. The same question as above, but assume the gravel packing is only 20 ft high.

4. Fit a 400-ft² house pad near the middle of a lot 30 x 100 ft, and design (per UPC) a septic tank and leachline (plus 100% expansion). Assume that the required absorption area is 50ft²/100 gal of septic tank capacity. The house has two bedrooms; the terrain is flat and has no trees.

5. The formulas derived in Chapter 13 have some limitations, even from the mathematical point of view. Under what theoretical soil conditions would they not be accurate and thus require iterative calculations?

6. A sanitarian measured the vacuum generated by septic tank pumper trucks within his jurisdiction. The results varied from a (suction head) low of 12 ft of water to a high of 27 ft of water. In the local mountain areas, some septic tanks were being installed at inaccessible locations; the truck vacuum pumps would be faced with lifting pumped sewage more than 50 feet up (*vertical* distance). No pumper used centrifugal pumps, and none was about to.

The sanitarian decided to forbid the installation of septic tanks at locations more than 25 (vertical) ft down from the assumed height of the truck vacuum pumps.

Was this a good decision? (The answer is at the end of this appendix.)

7. Some fine gravels might retain water and foul up the measurement of voids by the methodology presented in Chapter 6. How would you avoid this fouling up with only

two implements: a little drill to make a hole in the can, and your finger?

8. A seepage pit test hole set as in Problem 3 above is gravel packed to near the top. The top 3 ft is backfilled with native soil. Water is rapidly pumped from a water truck into the hole through the perforated 2-in. pipe. After a few minutes, and rumbling noises, water erupts 10 ft into the air from the top of the pit test hole (through the tip of the perforated pipe). Just as with Old Faithful, the eruptions are fairly timely for a while. But with time, the eruptions occur less frequently and reach lower heights. (Are you a practical joker? Try it out, you'll have fun!)
Explain this phenomenon.
Could the same thing happen in any gravel-packed hole?
Could rate-of-fall measurements be affected?

9. In your own words, prepare a list of dos and don'ts regarding use of septic systems.

10. Seepage pits were used in a new subdivision tract (40 half-acre lots, residential). The soils were generally coarse-textured, very permeable, and deep. Pits 15 ft deep have lasted well over 10 yr in the general area. All of a sudden, the pits in three lots failed simultaneously after only 3 yr of use. The septic system contractor checked his records and obtained samples of the liquid in the pits. This is his report:
a. The 17-ft-deep pits were excavated through (top to bottom) 12 ft of sand, followed by 2 ft of clay, followed by 3 ft of sand.
b. The liquid in the pit was not clear; it had black sludge particles.
Can you offer explanations for the failures? (Two hypotheses are offered at the end of this appendix.)

11. Two soils have identical percent voids, identical particle densities, and identical bulk densities. Can you assume their k's are identical? Why or why not?

12. Imagine an isotropic sand stratum 10 ft thick over completely impermeable bedrock. The k of the sand is 1 ft/day. The k' of the bedrock is zero. The terrain is horizontal and flat for miles. A 100-ft-long leachline discharges 1 ft³/day/ lineal ft. Assume flows obey Darcy's formula. Disregarding evapotranspiration and rainfall, a. How long would it take for the top of the mound to rise 8 ft above the bedrock? (Hint: the final formulas derived in Chapter 13 are useless when k' is zero. Make slightly different assumptions, and derive a different formula that will give you the dimensions of h and X after one day's Q has mounded. Assume the ratio of h to X will hold when the rise is 8 feet.)

b. What is the maximum rise of the mound if the terrain slope becomes 40% and the leachline follows contour lines?

c. What is the maximum rise of the mound if the terrain is horizontal and flat again but 3 ft of water is lost to evapotranspiration per year?

13. How would you go about estimating the (horizontal) k of an aquifer? (Many answers are possible.)

14. A large-diameter, 4-ft-long vertical tube is full of 3/4-in.-diameter gravel (high uniformity coefficient). For our purposes, the gravel has 33% voids. The tube is used to measure the k of this gravel. The result is that k = 40 ft/day. The measurement is repeated, but this time a tracer dye is uniformly mixed in with the water, at the top of the tube. Assume laminar flow.

a. Approximately how long would it take for the first coloration to be visible at the bottom of the tube?

b. Let us repeat the whole thing, but this time with a sand of low uniformity coefficient, 33% voids, and a k of 4 ft/day. Approximately how long would it take for the dye to appear at the bottom, especially in comparison to the gravel (i.e., would it take 10 times longer)? Why?

15. Forget about the Uniform Plumbing Code and explore in your imagination different shapes of excavations to create

absorption surfaces for sewage disposal: U shape, V shape, vertical cylinder, deep trench, etc. Figure out the least costly type of excavation that will make the best possible use of a "load" of gravel (a truck-trailer combination carries a load of about 22 tons or 350 ft³ [about $300 locally]). Assume you work with your local excavation costs, and your local gravel cost. Envision the following different situations:
   a. Root-plugging hazard
   b. Soil sloughing hazard
   c. Groundwater at 10 ft

16. In the pit falling head formula, L is the *arithmetic* average length of the water column. Is this a correct average? If the soil is isotropic, does the water column fall at a constant speed from a height of, say, 30 ft (above bottom) to a height of 15 ft? How would you compute the true value of L?

17. A consultant follows the UPC to the letter. He designs a septic system as follows:
   One hundred feet upslope from a neighbor's drinking-water well, the consultant installs a leachline. The soil is "coarse sand and gravel," and is loaded at the UPC's rate of 5 gal of sewage/ft² of absorption area/day. The perk time of the soil is 0.1 mpi. Groundwater is 5 ft below leachline bottom. When the neighbor's well is not pumping, the water moves downslope at a velocity of 30 ft/day.
   You are the sanitarian who has to approve the septic system. What will you do?

18. In a uniform soil, the measurements from 40 leachline test holes, 6 in. in diameter, average out to a normally distributed mean of 10 ± 1 mpi. If the testing will be replicated with 12-in. test holes, what mpi's might be measured?

19. In a uniform soil, the measurements from 40 seepage pit test holes, 6 in. in diameter and 30 ft deep, average out to a normally distributed Q of 20 ± 2 gal of water/ft²/day. If the

testing will be replicated with 12-in.-diameter holes, what Q's might be measured (per the falling-head formula)?

20. Use the formula to correct for gravel packing, and calculate the error from ignoring the thickness (volume) of the perforated pipe that holds the gravel in place. Assume that the inner diameter of the pipe is 1.5 in., and that the outer diameter is 1.75 in. Assume the test holes are 6 in. and 3 in. in diameter.

ANSWERS TO PROBLEMS 6 AND 10

6. If your answer is yes, congratulations. You are intelligent and well educated, and you know physics. One cannot lift water more than about 27 feet by vacuum.

However, your answer is absolutely incorrect. But don't let that bother you. You are in excellent company.

I was the sanitarian mentioned in Problem 6. I had drafted rules and regulations to prohibit installation of septic tanks too far down from a pumper truck's vertical reach.

One day I learned from a reliable source that the pumpers in our mountain area, who didn't know that it was impossible to do what they were doing, were in fact pumping septage up 40 to 80 feet by vacuum!

I went to the mountains, and found out that it was possible. The pumpers allowed air bubbles to be sucked up with the septage. As the bubbles rose, they pushed the fluid up. This was a slow process. Some pumpers accelerated the process and increased the lift power by introducing compressed air into the lower end of the hose.

(Fortunately, the rules and regulations I drafted got lost in "the paper shuffle," and never saw the lights of the Board of Supervisors' chamber.)

I tested Problem 6 on a couple of top-notch engineers. As I suspected, they also knew enough physics to assure me pumpers couldn't pump that high. (Some of us educated professionals would not qualify to be good septic tank pumpers!)

10. There are many possibilities. In my opinion, the most likely hypotheses are:

a. Septage plugged the bottom 3 ft of the pit. Water started mounding above the clay and created anaerobic conditions that promoted faster development of a lower-permeability biomat over the rest of the pit. The septage got out of the (two-compartment) septic tank because it was in suspension. This may happen when there is a thermal inversion and the bottom (sludge) layer rises to the top of the tank. Also, some septic tank additives (bacteria, enzymes, yeasts) generate little bubbles that rise as a froth and can attach to and move sludge particles.

b. The clay was smeared around the bottom 3 ft of the pit during pit excavation, and sealed it. Mounding, along with escaping sludge, did the rest.

Appendix M

AN INTRODUCTION TO THE WORLD OF LAW

What one does in professional practice depends not only on one's technical knowledge, but on legal and ethical constraints. Let us start by getting acquainted with our legal environment. Later on, in Appendix R, we'll delve into ethics.

I was first introduced to the influence of our legal/judicial system by my supervisor in 1974. Soon after I began my practice as a county septic systems specialist, I detected falsehoods and deficiencies in a report submitted by a developer. My supervisor told me to approve the report if it was signed by a registered civil engineer. He explained that liability would rest on the engineer, not on our county. That's all that mattered(!).

Many years later, an engineer (who read *Septic Systems Handbook*) properly designed leachlines along contours of sloping terrain without using a distribution box, in order to avoid premature failure. (Appropriately, he used drop boxes.) But the county sanitarian insisted that he use a distribution box, because the Uniform Plumbing Code (still) requires it. The engineer consented after the sanitarian stated his demand in writing. If the engineer is sued for designing a short-lived leachfield, he can rightfully blame the county. The county is probably immune to a lawsuit because the sanitarian followed regulations and "accepted practice." The property owner who paid for the leachfield is the one who may suffer the consequences.

Soon after the devastating 1988 Armenian earthquake, a story was in the news. Officials brought food and medical supplies to survivors in a devastated Armenian village. However, they refused to distribute such items to the survivors because there were no local officials that could sign the receipt forms: they had all perished in the earthquake. In a totalitarian state, strict and idiotic adherence to rules or regulations is quite understandable. But in "the land of the free and home of the brave," such events are also rather common, in part thanks to today's legal/judicial system originally designed to keep us free. So common are these instances that they are accepted as inevitable facts of life.

For instance, acting as a private consultant, I complained once to a county septic system specialist about an absurd requirement. Everything became clear in a few seconds as the specialist justified his position in front of my client and my tape recorder:

> I will explain to you what I was told when I started this job [as a septic systems specialist]. That I follow the standards as they are in the book. I was told that the previous specialist did things by policy, without approval from anybody, and that will not happen [here]. You [specialist] will go to us [departmental supervisors] and the Board of Supervisors and follow procedure. You won't have any problem if you follow procedure because your liability is zero. Consultants who follow procedure because the county dictates the procedure have no liability because the county dictates procedure. When you deviate, you as [a] consultant are liable for things that may fail prematurely.

In other words, he said that one must avoid the possibility of lawsuits and "go by the book" even if "the book" says that earth measurements are to be based on the earth being flat. He might have been absolutely correct, if avoiding liability takes precedence over serving the public. (I strongly disagree with such a position, particularly if there is substantial proof that the earth is spherical.) But, alas, he had a point.

The above-criticized way of thinking is not common only to public agencies' personnel. Private geologists and engineers have confided to me that sometimes they had to do what was legally safe (and followed irrational, counterproductive rules) instead of doing what was professionally honorable. I opened my eyes and ears to the news media and found out that the same problem is common in many other fields of endeavor. (Just recently, the news reported that a drastic increase in cesarean births, well beyond patients' needs and wishes, is due to obstetricians' attempts to lower the risk of lawsuits; *Forbes* magazine often carries stories about how our tort system is ruining our country economically.)

So, I decided to explore in some detail how "the law" works and influences everybody in general and how it impacts the field of septic systems. At first I started with armchair deductions and inferences, and then I compared these with the world I saw depicted in the media. (It is possible that what I saw in the media was biased by a need to vindicate my personal views. I don't think this was the case; but, you be the judge.)

The following sections will encompass armchair evaluations, comparisons with revelations in the media, some pertinent experiences with the law, and opinions and suggestions on how professionals can practice honorably and survive under our present legal/judicial system.

M.1 LAWS

Most laws originate in our legislative bodies. Major laws usually specify wishes or abstract goals and are defined by the judicial system at court trials.

It is generally believed that the purpose of all governmental activity is to induce "the greatest good to the greatest number of people." Therefore, one would expect that we would elect as our legislators people versed in ethics and economics, management science, or related fields. Judging from the news, it appears that we do not elect such enlightened people intentionally.

However, our representatives are well versed in passing laws. According to Perkins,[1] during 1988, 1600 new bills were approved in California; at the federal level, Congress passed over 2700 bills. Quoting Perkins, "with the . . . continuing deluge of new laws, you will find that the state agencies have switched from enforcing existing environmental programs to creating an incomprehensible plethora of special administrative and technical units that spend 90 percent of their time developing unrealistic standards and the other 10 percent trying to get local governments to enforce them. A sort of policy formulation without implementation." The resulting problems have ". . . caused Sacramento to pass additional legislation. This in turn has further confused those who are regulated as well as those who regulate and so on ad nauseam."[1]

ARCO's general counsel, Francis McCormack, echoes Perkins: "As a nation, we are over-legislated and over-regulated, burying ourselves in laws and litigation. Things have gotten so complicated that management can't move without talking to lawyers."[2]

Maybe the need for laws increases logarithmically with the lack of ethics of the population, among other things. A trillion laws can't force a people who are not ethically oriented to "do right." Each and every employee or businessman would have to be regulated and inspected, and the regulators and inspectors themselves would have to be regulated and inspected, and so on until the number of people required to regulate and inspect constitutes a sizable part of the population. And even then, who would regulate and inspect the topmost regulators and inspectors?

The burden of laws and regulations as a substitute for ethical foundations is astronomical. Let's assume that in a field of law there are only 99 laws enacted. If we add one additional law, how much time would it take to compare this law with the other ones, one, two, three, 100 at a time, in order to see if and how they relate to each other? Let us assume that each comparison lasts just one second and a lifetime lasts 70 years. The answer is that one would have to spend a full 250,000,000,000,000,000,000 lifetimes (without any sleep).* This gives some validity to a statement attributed to Winston Churchill (if I recall correctly): "When you have a thousand laws, you have no laws."

Though human brains can do a bit better with complex matters because they behave like "parallel processors" and process information according to selective cues, the previous astronomical number illustrates the complexity and problems involved when too many laws are enacted as a substitute for a few ethical rules.

The Incas are said to have managed with three laws ("don't lie, don't steal, work hard"), and the ancient Hebrews with 10. Perhaps their legislators were better at math and logic.

M.2 THE JUDICIAL SYSTEM

I am disturbed by what I see in our judicial system. The symptoms of its malaise are rather evident: the victims' rights movement, exposure of the "tide of lawyer misconduct, unethical behavior, and outright criminal activity,"[3] congested courts and lengthy trials, and the complaints of professionals in many fields that they have to pay immense insurance fees and practice not properly and ethically but in ways that avoid the mere possibility of a lawsuit.

It seems that the present judicial system imposes impossible tasks on its people. Take, for instance, a superior court

---

*The number of comparisons or combinations of n things, one to n at a time, is given by $Cn,n = 2^{n-1}$. Since 2^{10} is approximately 1000, it follows that 2^{100} translates into roughly 1,000,000,000,000,000,000,000,000,000,000 seconds, and 2^{99}, half as much. So, an additional law would require 500×10^{27} seconds. A lifetime of 70 years has a total of 2.2×10^9 seconds. Hence one needs 250×10^{18} lifetimes.

judge. He must rule alone over a trial in a fair, alert, and impartial way, even if his personal problems may affect him the day of a trial. He is supposed to understand every subject imaginable, although this is impossible in our specialized, technical world. As help, he has counsels for plaintiff and defendant; but both counsels are free (if not encouraged) to zealously do their utmost to deceive judge and jury. Surgeons, who often deal with life-and-death situations, receive rigorously supervised tutelage and specialized training before they are allowed to practice on their own. Judges deal with issues that are even more grave, yet their training is not as rigorous, and there are no judge specializations (say, real estate, contracts, etc.)

As another case in point, the adversary system allegedly is "the best way to arrive at the truth." In medieval Europe, it was taken for granted that dueling between adversaries would elucidate who was the righteous disputant, regardless of fighting ability. (Divine providence would help the good guy win.) This absurd, grotesque tradition still survives in our judicial system (and without the excuse or pretense of divine intervention): The outcome of a trial often depends more on a lawyer's ability, trickery, or lack of scruples than on the merit of his/her client's claims. And since notoriously successful lawyers charge very high fees, the corollary is that the outcome of a trial often reflects litigants' income disparity. "Justice" that can be bought violates every ethical and moral code I know of.

The prominent trial attorney and law professor A. Dershowitz was confronted by G. Rivera's question, "Do you deny, sir, that there is an unequal system of justice in this country and that a rich man can buy an acquittal?"[4] Though somewhat extreme, his reply was, "I do not deny it. I proclaim it. It is one of the worst aspects of our criminal justice system. Justice in this country is for sale, it's for rent, and it's for barter."[4] He should know.

Last year I tried to figure out what's happening with our legal system by reading a few issues of *California Lawyer*, a magazine published by the State Bar of California. I tried to find articles about subjects fundamental to any large modern enterprise, say, queue analysis to optimize court load distribution, or critical paths to streamline legal procedures, or cost effectiveness and cost efficiency, or instituting quality control mechanisms, or improving evaluation and feedback mechanisms so legislators know how well their laws are working, or challenges to (or validation of) the fundamental premises on which the judicial system is based (like, for instance, the premise that the adversary system is "the best way to arrive at the truth.") I found nothing, absolutely nothing of the kind. (I remember that a big deal about "innovation" assumed the form of an article describing a solution to a local shortage of judges: lawyers acted in their stead. A "Band-Aid" cure for cancer!) And I recall that in those articles I found little or no criticisms of, or by, the legal profession. The mentality that spoke to me from the pages I read was that of smart people of the 18th century, before the scientific revolution changed our way of reasoning and analyzing things.

However, I was delighted when I found out that the *Daily Journal*, apparently the largest legal journal in the nation, did publish soul-searching articles. In a semi-random way, I read two issues per week between February and March, 1989. Though still within the mindset of the 18th-century world, their lively articles offered candid critical views that I will share with my readers. At about the same time I found provocative discussions about the judicial system on the TV channels. From all these sources, I gleaned statements by people who should know very well what the shortcomings of the judicial system are.

One of these people is Michael Josephson, founder of the Josephson Institute for the Study and Teaching of Ethical Choices. Bill Moyers began a television interview with Josephson[5] by contrasting Josephson's philosophy with that enunciated by J. R. Ewing ("Once integrity goes, everything else is just a piece of cake"). Josephson taught law and ethics. In the ethics course, his original emphasis was on how to avoid rules' restrictions. But after the birth of his son, he wondered what kind of world he would bequeath to the child and changed his outlook drastically. He said, "I have been a law professor for the last 20 years. I certainly understand the law and the rules. And one of the things I understand is that you can't write a law I can't get around." As he sees it, "the underlying purpose of the law is a social statement, and this should take precedence over the mechanics of the law."* Otherwise, we have a "dog-eat-dog society in which everybody is lawyering everybody else, pushing the rules."[5] He pointed out that ethics consists of caring about other people, that our society is at present rights-oriented rather than responsibility-oriented, and that nowadays personal motivation is geared toward competition and victory rather than doing the right thing.

At the American Bar Association 1988 convention in Denver, Archibald Cox delivered a speech (see the *Daily Journal* [Los Angeles] of February 16, 1989) that brought the symptomatology of the problem into focus. Excerpts follow:

> Exclusive loyalty to a client is sometimes inconsistent with the legal system or the public good. In litigation, endless discovery, mountains of documents, and frivolous appeals that may produce a favorable settlement or even win the case will often be incompatible with the fair and reasoned administration of justice. . . .
>
> We [lawyers] have largely accepted . . . the system of administering justice devised for the 18th and 19th centuries and contented ourselves with perfecting the details of the old machine. Both fair minded observers and much of the public regard the whole system of administering civil and criminal justice as slow, creaky, haphazard, and incredibly expensive.
>
> When substantial reforms are proposed, does not the public position of segments of the organized Bar often correspond to the selfish interest of lawyers or their clients as to

---

*The same thing was enunciated almost 2000 years ago: "The Sabbath (the law) was made for man, not man for the Sabbath." That is, one should follow the spirit rather than the sacred letter of the law.

raise public criticism about the . . . profession's disinterested dedication to the fair and efficient administration of justice?

Often, we seek to hide the conflict, stressing the duty of loyalty to our clients, and then rationalize preferring the client's interest over concern for the integrity of the law by saying that justice is best served by the adversary system.

Some of the adversary system's flaws were revealed in the PBS series "Ethics in America." Excerpts follow. For clarity's sake, most of my personal notes or comments are in italics.

During the broadcast of a panel discussion[6] I heard this exchange:

> From a New York University law professor: "Lawyers, especially criminal defense lawyers, are amoral agents that can justify their behavior only by being part of an adversary system that is based largely on the Constitution and that is amoral. But we can change it. They [*lawyers*] are told to behave this way."
>
> A Massachusetts district attorney added, "The constitution governs these particular issues."
>
> R. Duncan, judge (Superior Court of Alameda County, California), interjected (with what I felt was a tinge of annoyance in his voice): "It is lawyers that made such rules for lawyers. And the public and the legislature get a chance to look at it, but basically nobody is paying any attention to lawyers who debate these issues at conventions. But it is only they who are doing it. It isn't society."

Another discussion[7] featured very prominent panelists. Attorney J. F. Neal assumed the role of a defense attorney. He admitted that, at trial, knowing full well that the prosecution's witness was telling the truth, he would still try to make this witness look like a liar. He said that in the adversary system, "Out of the clash will come truth and justice." J. Smith, philosophy and ethics professor at Yale University, retorted that the only thing he ever saw coming from adversary confrontations was conflict. Neal admitted, "My goal is to deceive the jury. But that is permitted by the system."

Smith replied, "Then we must take another look at the system." U.S Supreme Court Justice Antonin Scalia tried to defend the system, stating, "By and large it works well." [*In my opinion, this is a value judgment without any scientific basis whatsoever. In comparison to some other countries' systems, he might be right. In comparison to a hypothetical but feasible system with scientifically defined objectives, with built-in feedback evaluation and other quality control mechanisms, and ultimately cost efficient and cost effective, the available evidence is that it is working very, very poorly*]. Justice Scalia further postulated that the alternative is an inquisitorial system with investigative judges: "And then you win or lose depending on the judge. [*In the adversary system*] if you hire a bad attorney, it is your fault." And how is the average person going to know in advance whether his attorney will turn out to be bad or not? Furthermore, municipal court judges can and do perform as investigative judges.

Near the end of the panel discussion Neal retracted his previous statement about the adversaries' clash engendering truth and justice. He stated, "Don't put me in a position that I have said that I am out for the truth. The adversary system is not calculated to produce the truth, but justice." He said this earnestly; he wasn't joking or quoting something from Orwell's *Animal Farm*. His statement points out what's wrong with the adversary system in a most striking way!

M.3 LAWYERS

If justice is the glue that holds society together, lawyers are the principal glue-makers. Lawyers speak for the legislature. They might not author laws, but they write and interpret laws. At court trials, they define and refine laws. They discipline all other professions and trades. And (unfortunately, in my opinion) they have been given the privilege of disciplining themselves, as if they were ethical paragons.

I know about a dozen lawyers well enough to venture an

opinion about them. Some are extremely honest and decent. (One of them told me he was so disgusted with his successes at court trials defending people he knew were guilty, and also about what he perceived to be an unjust judicial system, that he quit this line of work, became a public employee, and engaged in activities that help society.) As for others, I would not call them swine, as such comparison would be unfair and highly offensive to swine. The rest fall anywhere between the two extremes.

But I would not dream of extrapolating characteristics from a small or even a large sample of lawyers to the whole of the profession. It is meaningless to ask, "Are lawyers crooked?" Crooks can be found in every profession. The proper questions to ask are two. The first is, "Is a profession's code of ethics truly ethical, the way ethics is understood by philosophers and common people?" And the other one is, "Does a profession have good mechanisms to remove its incompetent and crooked members?"

In the case of lawyers, the answer to the first question appears to be, "Not nowadays." Ethics is concerned with the greater good of society, not just the benefit of a lawyer's client. In the early 1930s, F. Wellman[8] noted how things used to be (in the late 1800s to early 1900s) when lawyers were regarded as highly ethical people, devoted to the public good. As he puts it, the famous Elihu Root advised some of his clients in this manner: "Yes, the law lets you do it. But don't: it's a rotten thing to do." And they would heed his advice.

Today things are so different that, according to Hall,[9] the State Bar of California contemplated a voluntary (!) "Code of Professional Courtesy" which includes statements such as, "I will remember that my responsibilities as a lawyer include a commitment not only to my client but to the public good," and "I will advise my client against pursuing litigation . . . that does not have merit." (!!!)

The second question has been answered by Walters and by Bumsted and Guttman. Walters[10] noted that the California Bar has a dual role: It regulates errant lawyers and also is a private lobby for lawyer's political interests (in plain English, it is the fox guarding the chicken coop). In his words, "Something has to give; that something has been the public interest." Bumsted and Guttman (Gannett News Service) wrote an expose[3] in serial form. For brevity's sake, I quote verbatim the series' daily abstracts:

Sunday: The nation's bar, bursting with an unprecedented number of lawyers, is failing to stem the tide of misconduct, unethical behavior, and outright criminal activity.
Monday: Lawyers stealing from their clients are a growing national disgrace: Theft is so widespread that at least 45 state bars or high courts have special funds to reimburse victims.
Tuesday: Leniency is the letter of the law in lawyer discipline. Few misconduct cases are pursued. Those that are result in penalties which tend to be lenient, late, and wildly inconsistent.
Wednesday: The nation's self-regulated bar fails to protect the public when it hides complaints against lawyers and prosecutes misconduct in secret, closed hearings.
Thursday: Critics within the profession warn that lawyer privilege of self-regulation could be jeopardized unless the bar improves discipline.

From the title of an article in the *Daily Journal*,[11] "Forget what Shakespeare said" (which is, "Let's kill all the lawyers"), I surmise that its author is sensitive about lawyers' public image. Its author points out that it wasn't Shakespeare who didn't appreciate lawyers, but one of the despicable characters in *King Henry VI*. After reading his article, methinks that this Shakespearean interpreter doth protest too much about the negative perceptions regarding lawyers being this or that. (I do not make much of these protestations because—and I trust my enlightened readers do likewise—I judge not all lawyers by what some do or are.) Also, two of his statements are symptomatic of a contagious form of legal naivete: "American citizens are not necessarily more litigious than other peoples: they are just more often put in situations where disputes are likely to arise." And, "Our system does not work perfectly." (Note that this last sentence can be used by the Gulag's bosses to justify their system, too.)

The *National Law Journal* was quoted re the attendance at the recent California Bar convention, in Monterey, 1988. As reported,[12] overflow crowds attended the "ever-profitable" session entitled "Current developments in estate planning, trust, and probate law." The session entitled "Why are they saying all those terrible things about lawyers?" drew a "fair crowd" but "most left before the session ended." The "least popular" session "drew barely a dozen people at any one time" and was entitled, "Practicing ethically and successfully." Also, it has been reported that a committee of the American Bar Association offered a resolution opposing the regulation of the legal profession by executive or legislative branches of government at any level.[13] (I heard that it passed with unanimous approval.) From these reports in references 12 and 13, I infer that the Shakespearean interpreter mentioned in the paragraph above is not alone, whatever galaxy he might inhabit.

M.4 JUDGES

I know only one (Superior Court) judge well enough to form an opinion, and it is high. He sees judges as the "shock absorbers" (or people in the middle) between the legislature and the vast unmet needs of the population. He told me that sometimes he gets depressed by the way the judicial system is working—or not working.

I have asked lawyers about judges. I heard (in two different jurisdictions) that about 10% of the judges are incompetent, 10% are highly qualified, and the rest fall somewhere in between: Just a normal curve, though I wished it were biased towards excellence. As for judges' apparent callousness, an attorney told me that he knew a judge who really cared and visibly aged 20 years in his first three years on the job: wrinkles, white hair, always very serious. Judges' responsibilities are enormous; I suppose that if they do not become detached from the tragedies they see in court, they are at risk of wrecking their own lives.

Rheingold offered a trial attorney's perspective about judges[14] (and, in so doing, a perspective on himself). According to him, though most judges are good, there are problem judges who "threaten to thwart not only the case but the fairness of the judicial system." He classified the types of "problem judges" (and recommended strategies to cope with them): arrogant ("most common" fault); dumb; too-smart ("they know all the answers and do not want to listen to you"); intemperate; biased against you; biased for you ("they might make the jury feel sorry for the other side"); indecisive; lazy; rushed. The strategy he recommended to cope with arrogant judges is to suffer them and "tell yourself that he didn't have a successful practice [*as a lawyer*], which is how he got on the bench."*

Judges are supposed to follow precedent so that the law will not be erratic and capricious. But history shows that honors accrue to judges who break with precedent (Brandeis, Warren). Verdicts in higher courts (appellate, state and federal Supreme Courts) prove that judges tend to be fallible in direct proportion to an appellant's ability to carry the litigation to ever-higher courts.

On one hand judges have much power; on the other hand this power is restricted in ways that may violate true (ethical) justice. For instance, late in 1988 the news referred to a case in which a person was prosecuted for manufacturing an illegal "designer drug," a derivative of methamphetamine. The judge threw the case out because the law as written had merely banned a non existing drug, "ethamphetamine" derivative: the "m" was missing due to a simple typographical error. I hypothesized two alternative reasons for the

---

*Could it be that such a lawyer was too ethical to be "successful"? As a former public employee, I strongly disagree with Rheingold. I suppose that some people whose main concern is making money might find it difficult to understand that many people go into public service just because they do wish to serve the public.

judge's actions: One, the judge decided that the intent of the legislators who enacted the law was not ascertainable, on the axiom that legislators so stupid (as to allow a judicial system to run the way it does) could not be trusted to know what they meant when they enacted the law. Or, alternatively, the judge feared a reversal on appeal if he had not followed the letter of the law precisely. Lawyers told me that the second hypothesis was correct. (This validates Josephson's views; see Section M.2 in this Appendix.)

Judges generally campaign for election and reelection. Unfortunately, their campaign costs are sometimes extremely high. Some judges have accepted bribes and have repaid campaign contributors with legal favors.[3] In "Justice for Sale" Methvin tells us how "a handful of millionaire lawyers reshaped the Texas Supreme Court to their liking" with money and favors, and how the people of Texas "won it back."[15]

It has been reported[16] that last year a record number of California judges were disciplined. California has 1462 judges. 693 complaints against judges were reviewed by the Judicial Performance Commission. But what do these figures mean? The subject report does not explain. Too many possibilities obscure their meaning, for instance:

- There are too many unqualified judges.
- Excessive judicial workloads are taking their toll.
- Californians have a propensity to complain.
- Mechanisms to discipline judges are working well.
- A few judges might account for most of the complaints.

M.5 THE JURY

What can be fairer than a representative cross section of a community, knowledgeable of the facts and free from bias? Nothing, in my opinion. However, our judicial system permits lawyers to practice deception, and sometimes does not permit a jury to know all the facts. As a case in point, Alcee

Hastings is the federal judge who was accused of bribery and lying under oath, and was acquitted by a jury, apparently because much of the evidence was not admitted at his trial.[17] But Congress is not bound by court rules. The House examined all of the evidence, and impeached him by a vote of 413 to 3.[17]* Of course, the judge claims that a jury absolved him and that the legislators should not tamper with the jury system.[17]**

M.6 COURT TRIALS

The dictionary defines the word "trial" as "hardship, pain, source of annoyance or irritation." Perhaps excepting some trial attorneys (for whom trials are a source of income), this definition appears to be rather accurate.

Words are not enough to describe how the courts work. A fascinating view can be obtained by watching the TV program "People's Court." It and its judge have been criticized by a U.S. Appellate Court judge.[18] But I found "People's Court" true to reality (as I have seen it myself as a plaintiff and as an expert witness). I'd strongly recommend watching this program daily for at least a month. Then envision what may happen in a Superior Court thusly: Imagine that one (or both) of the litigants does not speak his mind freely (as litigants in "People's Court" do), but has a lawyer who may facilitate the appearance of false witnesses,*** may coach on

---

*On October 20, 1989, the press reported that the Senate followed suit.

**Wouldn't the American Bar Association resolution against interference from the legislative or executive branches of government[12] support his position and implicitly endorse corruption of the judiciary and hence of the whole legal system?

***When I noticed that proven perjuries were not punished in a California Superior Court series of trials, I asked why, and an attorney told me that it's because judges assume that everybody lies. (So people find out that being truthful is no advantage and are not afraid of perjuring themselves.) He also said that in federal courts perjury is punished. I checked with other lawyers; they confirmed his statements but noted that on rare occasions a local judge will punish a perjurer to "send the message out."

how to lie under oath or plead the fifth Amendment, may try to suppress the evidence from the other side, and/or may try to delay the trial incessantly to wear down the other litigant. (The other litigant may die or run out of money, or his witnesses may die, move away, or forget the details of what they were to testify about.)

I wonder what the Pledge of Allegiance (". . . and justice for all") means to the people within our judicial/legal system. I think it was former Chief Supreme Court Justice Warren Burger who said that our system of justice was unworthy of a civilized country.

All in all, no one is safe.

Despite the bleak picture presented so far, there is no cause for despair. In subsequent sections we'll see that there is hope for the future, and we'll also see that engineers, geologists, sanitarians, or other private consultants may practice honorably with little fear that a lawsuit will victimize them, as there is an "alternative judicial system."

M.7 HOPE FOR THE FUTURE

In California, Senator R. Presley authored a bill in 1984 authorizing the teaching of ethics and civic values in California schools. The superintendent of public instruction (William Honig) procured funding, and, according to a school vice-principal,[19] such teaching is now widespread in California.

On June 21, 1989, the ABC Evening News with Peter Jennings carried a very encouraging story. Some schools across the nation are teaching "old values" (honesty, truthfulness, caring about other people, honor, etc.) again in special courses. And surprise! Behavior and grades improved noticeably throughout those schools. In the 1960s schools quit teaching values (out of fear that someone might sue), but now at last it is "in" again to inculcate ethics and civic values.

At another level, worthy of imitation in other states, Senator Presley authored landmark legislation in 1988 (SB 1498) to reform the State Bar attorney discipline system. Laws have been enacted to provide for a State Bar Discipline Monitor and to provide professional administrative law judges to the State Bar. The Monitor reports to the legislature (and to the public) how well the lawyer discipline process is working. Professional administrative law judges are replacing the volunteer lawyers who until recently conducted disciplinary hearings against their errant colleagues.

Another cause for optimism is that undoubtedly our legal/judicial system will be brought up to the 20th century. It's inevitable, even if it takes 40 centuries.

Individuals who can't wait 40 centuries or so may wish to speed up the process and help themselves by joining and/or using the help offered by the organizations mentioned in Appendix N.

REFERENCES

1. Perkins, K. 1989. California environmental policy: a time for change. *Calif. J. Environ. Health* 12:16–18.
2. *Daily Journal* (Los Angeles), March 29, 1989 (California Law Business section, p. 9).
3. Bumsted, B. and J. Guttman. October 19–23, 1986. Beyond the law. *The Sun* (San Bernardino County).
4. "Geraldo: The best defense that money can buy." CBS broadcast of January 25, 1989. Transcript obtainable from Journal Graphics, 267 Broadway, 3rd Floor, New York, NY 10007.
5. "Bill Moyers' World of Ideas: Interview with Michael Josephson." PBS broadcast of September 14, 1988. Transcript obtainable from Journal Graphics, 267 Broadway, 3rd Floor, New York, NY 10007.
6. "Ethics in America: To defend a killer." PBS broadcast of February 7, 1989. Videotape obtainable by calling 1–800

LEARNER or contacting The Annenberg CPB Collection, 2040 Alameda Padre Serra, P.O. Box 4069, Santa Barbara, CA 93140.

7. "Ethics in America: Truth on trial." PBS broadcast of March 21, 1989. Videotape obtainable by calling 1–800 LEARNER or contacting The Annenberg CPB Collection, 2040 Alameda Padre Serra, P.O. Box 4069, Santa Barbara, CA 93140.

8. Wellman, F. L. 1932. *The Art of Cross-Examination*. Republished in 1986 by Dorset Press, New York, NY.

9. Hall, M. 1989. State bar wants return of courtesy to the legal profession. *Daily Journal* (Los Angeles), March 6, p. 6.

10. Walters, D. 1989. Lack of oversight is destructive of the state bar (reprinted from the *Sacramento Bee*). *Daily Journal* (Los Angeles), March 6, p. 6.

11. Moskowitz, M. 1989. Forget what Shakespeare said. *Daily Journal* (Los Angeles), January 18, p. 6.

12. Briefs by Amicus. 1989. California Bar: What's in, out. *Legal Reformer* 9:28.

13. Hall, M. 1989. Hard issues set for ABA House at winter meet. *Daily Journal* (Los Angeles), February 2, pp. 1, 24.

14. Rheingold, P. D. 1988. How to cope with problem judges. *Daily Journal* (Los Angeles), February 14, 1989, p. 6 (reprinted from the November 1988 issue of *Trial*, published by the Association of Trial Lawyers of America).

15. Methvin, E. 1989. Justice for sale. *Reader's Digest*, May, pp. 131–136.

16. Hager, P. 1989. Record number of judges investigated, disciplined. *Los Angeles Times*, February 25, part I, p. 35.

17. Robinson, M. 1989. Judge asks impeachment halt. *Daily Journal* (Los Angeles), March 16, p. 2.

18. Mikva, A. 1989. The verdict on Judge Wapner. *TV Guide*, April 22, pp. 12–14.

19. S. Fleischer, personal communication, July 15, 1989.

Appendix N

JOHN Q. PROFESSIONAL VERSUS THE LEGAL SYSTEM: HELPFUL ORGANIZATIONS AND "ALTERNATIVE JUSTICE"

Many organizations are trying to reform the legal system and/or help its victims or prospective victims. HALT is a nonprofit, tax-exempt organization.* With over 150,000 active members nationwide, HALT describes itself as America's largest organization for legal reform. I quote their meritorious goals as stated in their literature:

- Educate consumers about the law so that they can handle their own legal matters or work knowledgeably with a lawyer.
- Simplify legal procedures to make them comprehensible and accessible to the average citizen.
- Increase access to affordable legal services by increasing the range of legal services provided by competent nonlawyers.
- Promote the use of alternative methods of resolving legal disputes such as mediation and arbitration.

HALT publishes a plethora of self-help manuals. For instance, a mini-manual entitled "Going it Alone in Court" explains how to do just that, and offers sample forms, pleadings, and flow charts. "Using a Lawyer" provides guidance on how to select a lawyer, negotiate with one, work with the

---

*1319 F Street N.W., Suite 300, Washington, DC 20004

chosen one, and part. Another publication is entitled "Directory of Lawyers who Sue Lawyers" and offers advice on weighing the merits of one's malpractice claim, avoiding the need for malpractice suits, choosing a good malpractice attorney (it is very difficult to find lawyers who dare sue other lawyers), etc. Also, upon request, HALT may refer a person to lawyers sympathetic to its goals.*

The American Arbitration Association** offers a way out of the world of lawsuits. This nonprofit private organization was founded in 1926, when it became apparent that the judicial system "was not the best way of solving all types of disputes." Through the AAA, one may solve disputes without the problems common to court trials: high legal expenses, long legal delays, harassment from the other party, judges or juries who do not understand very technical matters and might be misled by the other party's lawyer or expert witnesses, etc. From the start, one has a good idea of how much the arbitration will cost (fees vary from about $300 minimum for small claims to about $14,000 for a $5,000,000 claim). Also, one knows that the process will be relatively speedy. The cost of a court trial is often unpredictable, moneywise and timewise. Contractors, consultants, and other professionals would do well to acquaint themselves with the AAA, its philosophy, and its procedures for arbitration and mediation. As for myself, after I found out about the AAA, I begun placing in some of my professional contracts a standard clause that takes any possible claim against me out of the realm of court trials and places it in the realm of expert-assisted AAA arbitrators: "Any controversy or claim arising out of or relating to this contract or any breach thereof shall be settled in accordance with the

---

*HALT's Director of Development, B. Alford, told me that out of 150,000–180,000 active members, 485 are lawyers (personal communication, 3/22/89). Interesting statistic. This is about one per 340 members, within the ballpark of the national ratio of one lawyer per 360 people. On the other hand, 485 lawyer-members out of over 700,000 lawyers nationwide might mean that less than 0.07% of lawyers know and care enough about HALT's goals to join this organization.
**140 West 51st Street, New York, NY 10020–1203

Rules of the American Arbitration Association, and judgment upon the award may be entered in any court having jurisdiction thereof."

If someone were to bring me to court, the California Code of Civil Procedures, Section 1281.2, would usually require the court to order arbitration per the clause above. Other states have similar laws.

Also, it should be noted that the parties to a contract may designate any arbitrator or mediator they feel comfortable with, not just the AAA.

Local offices of the AAA are listed below:

Arizona

3033 North Central Avenue,
   Suite 608
Phoenix, AZ 85012–2803
(602) 234–0950/230–2151 (Fax)
Deborah A. Krell

California

443 Shatto Place
Los Angeles, CA 90020–0994
(213) 383–6516/386–2251 (Fax)
Jerrold L. Murase

525 C Street, Suite 400
San Diego, CA 92101–5278
(619) 239–3051/239–3807 (Fax)

417 Montgomery Street
San Francisco, CA 94104–1113
(415) 981–3901/781–8426 (Fax)
Charles A. Cooper

Colorado

1775 Sherman Street, Suite
   1717
Denver, CO 80203–4318
(303) 831–0823/832–3626 (Fax)
Mark Appel

Connecticut

Two Hartford Square West
Hartford, CT (06106–1943)
(203) 278–5000/246–8442 (Fax)
Karen M. Jalkut

District of Columbia

1730 Rhode Island Avenue,
   N.W., Suite 509
Washington, DC 20036–3169
(202) 296–8510/872–9574 (Fax)
Garylee Cox

Florida

99 Southeast Fifth Street, Suite 200
Miami, FL 33131–2501
(305) 358–7777/358–4931 (Fax)
Rene Grafals

Georgia

1360 Peachtree Street, N.E., Suite 270
Atlanta, GA 30309–3214
(404) 872–3022/881–1134 (fax)
India Johnson

Hawaii

810 Richards Street, Suite 641
Honolulu, HI 96813–4728
(808) 531–0541/533–2306 (Fax)
Keith W. Hunter

Illinois

205 West Wacker Drive, Suite 1100
Chicago, IL 60606–1212
(312) 346–2282/346–0135 (Fax)
David Scott Carfello

Louisiana

650 Poydras Street, Suite 2035
New Orleans, LA 70130–6101
(504) 522–8781/561–8041 (Fax)
Ann Peterson

Massachusetts

230 Congress Street
Boston, MA 02110–2409
(617) 451–6600/451–0763 (Fax)
Richard M. Reilly

Michigan

Ten Oak Hollow, Suite 170
Southfield, MI 48034–7405
(313) 352–5500/352–3147 (Fax)
Mary A. Bedikian

Minnesota

514 Nicollet Mall, Suite 670
Minneapolis, MN 55402–1092
(612) 332–6545/342–2334 (Fax)
James R. Deye

Missouri

1101 Walnut Street, Suite 903
Kansas City, MO 64106–2110
(816) 221–6401/471–5264 (Fax)
Lori A. Madden

One Mercantile Center, Suite 2512
St. Louis, MO 63101–1643
(314) 621–7175/621–3730 (Fax)
Neil Moldenhauer

New York

585 Stewart Avenue, Suite 302
Garden City, NY 11530–4789
(516) 222–1660/745–6447 (Fax)
Mark A. Resnick

140 West 51st Street
New York, NY 10020–1203
(212) 484–4000/765–4874 (Fax)
Carolyn M. Penna

State Tower Building, Room
    720
Syracuse, NY 13202–1838 (315)
    472–5483/472–0966 (Fax)
Deborah A. Brown

34 South Broadway
White Plains, NY 10601–4485
(914) 946–1119/946–2661 (Fax)
Marion J. Zinman

New Jersey

265 Davidson Avenue, Suite
    140
Somerset, NJ 08873–4002
(201) 560–9560/560–8850 (Fax)
Richard Naimark

North Carolina

7301 Carmel Executive Park,
    Suite 110
Charlotte, NC 28827–8297
(704) 541–1367/542–7287 (Fax)
Mark Sholander

Ohio

441 Vine Street, Suite 3308
Cincinnati, OH 45202–2809
(513) 241–8434/241–8436 (Fax)
Philip S. Thompson

1127 Euclid Avenue, Suite 875
Cleveland, OH 44115–1632
(216) 241–4741/241–8584 (Fax)
Earle C. Brown

Pennsylvania

230 South Broad Street
Philadelphia, PA 19102–4121
(215) 732–5620/732–5002 (Fax)
James L. Marchese

Four Gateway Center, Room
    221
Pittsburgh, PA 15222–1207
(412) 261–3617/261–6055 (Fax)
John F. Schano

Puerto Rico

Esquire Building, Suite 800
San Juan, PR 00918–3628
(809) 764–8515/764–8848 (Fax)
Lillian A. Reyes

Tennessee

162 Fourth Avenue North,
    Suite 106
Nashville, TN 37219–2412
(615) 256–5857/256–7819 (Fax)
Tony Dalton

Texas

Two Galleria Tower, Suite
   1440
Dallas, TX 75240–6620
(214) 702–8222/490–9008 (Fax)
Helmut O. Wolff

1331 Lamar Street, Suite 1459
Houston, TX 77010–3025
(713) 739–1302
Therese Tilley

Washington

811 First Avenue, Suite 200
Seattle, WA 98104–1455
(206) 622–6435/343–5679 (Fax)
Neal M. Blacker

Appendix O

NEGLIGENCE IS THE LEGAL TERM FOR A PUNISHABLE GOOF

When professionals are sued for an alleged goof, the trial judge gives standard instructions to the jury on how to make a determination of guilt or innocence. I reproduce the (verbatim) instructions given in California, as found in the *Book of Approved Jury Instructions*, or BAJI.* (Instructions in other states are similar, but may not be identical.) The blank spaces are filled by inserting "engineer," "geologist," or similar professions.

BAJI 6.37:
In performing professional services for a client, a _____ has the duty to have that degree of learning and skill ordinarily possessed by reputable _____, practicing in the same or a similar locality and under similar circumstances.

It is his or her further duty to use the care and skill ordinarily used in like cases by reputable members of his or her profession practicing in the same or a similar locality under similar circumstances, and to use reasonable diligence and his or her best judgment in the exercise of his professional skill and in the application of his learning, in an effort to accomplish the purpose for which he or she was employed.

A failure to fulfill any such duty is negligence.

---

*The excerpts from BAJI are reprinted with permission from *California Jury Instructions, Civil,* copyright 1986 by West Publishing Co., St. Paul, MN.

BAJI 6.37.1:

It is the duty of a _____, who holds himself or herself out as a specialist in a particular field of _____, to have the knowledge and skill ordinarily possessed, and to use the care and skill ordinarily used, by reputable specialists practicing in the same field and in the same or a similar locality and under similar circumstances.

A failure to fulfill any such duty is negligence.

BAJI 6.37.2:

A _____ is not necessarily negligent because he errs in judgment or because his efforts prove unsuccessful. Such a person is negligent if the error in judgment or lack of success is due to a failure to perform any of the duties as defined in these instructions. [*The judge would instruct the jury as to what the duties are. The duties outlined in local manuals re perk testing/ site evaluation would probably be included.*]

BAJI 6.37.3:

[*This instruction applies only if it is included within the professional standards of the profession involved.*] Once a _____ has undertaken to serve a client, the employment and duty as a _____ continues until ended by consent or request of the client or the _____ withdraws from the employment, if it does not unduly jeopardize the interest of the client, after giving the client notice and a reasonable opportunity to employ another _____ or the matter for which the person was employed has been concluded.

BAJI 6.37.4:

You must determine the standard of professional learning, skill and care required of the defendant only from the opinions of the _____ [*including the defendant*] who have testified as expert witnesses as to such standard.

You should consider each such opinion and should weigh the qualifications of the witness and the reasons given for his or her opinion. Give each opinion the weight to which you deem it entitled.

You must resolve any conflict in the testimony of the witnesses by weighing each of the opinions expressed against the others, taking into consideration the reasons given for the opinion, the facts relied upon by the witness and the relative credibility, special knowledge, skill, experience, training and education of the witness.

From all the instructions above, one may infer that:

1. It is legally safe to goof if one can prove to a jury that other reputable professionals are just as guilty of the same type of error.
2. Bad judgment and bad results are not necessarily punishable unless failure to perform specified duties is involved. (Compliance with requirements in a pertinent "procedures manual," even if the manual's procedures are wrong, does much to protect professionals from liability.) The other side of the coin is that it may be dangerous to be innovative and strive for progress through one's work.
3. Solely on the basis of the defendant's and expert witnesses' opinions (often about complicated subjects, which jurors may not fully understand), the jury must determine whether a professional is guilty of negligence. And a key to this determination is relative credibility. (Unfortunately, from what I have seen and heard, it seems that lawyers can produce expert witnesses who will testify to anything; I heard one lawyer brag about his ability to procure such expert witnesses. In contrast, I understand that in Germany it is the judge who hires and pays the expert witnesses so they'll act as his truthful and objective guides.)

Appendix P

TESTIFYING IN COURT AND GIVING EXPERT TESTIMONY

It is possible that many of us will testify in court as expert witnesses, plaintiffs, or defendants. Knowledge of the field of battle is crucial. A thorough and concise map of the field has been published by Sanford M. Brown[1] and is reproduced at the end of this Appendix. Case histories in Appendix Q shed more light on the legal battlefield as well as on technical septic system matters. Below, I have written some observations to make Brown's map more complete. Additional insights may be obtained from Matson's book.[2]

GENERAL OBSERVATIONS

Before appearing in court, one should be aware of what I would call "the Frynne effect" (pronounced "free" as in freedom, and "ne" as in never). In Greece, during the classical period, a very beautiful woman called Frynne was charged with a crime. When she appeared in front of the judges at her trial, her attorney removed her tunic with a stroke of his hand and exhibited her, stark naked. "Could this woman be guilty?" he asked the judges. The judges saw the exquisite evidence and concluded to a man that she was not guilty. The Greeks of the period (wise Socrates included) "knew" that Beauty is Virtue, and Virtue is Beauty. Well, even today, when we see movies, in the first scenes we judge who the good guys are just from their good looks. So do jurors at trial, subconsciously. Of course, nowadays it's not how well we look when we are undressed, but when dressed. Appropriate clothes (like those worn by respectable local lawyers), and conviction in one's "body language" and in one's voice add up to credibility. Appearances do count.*

A witness must beware of hypothetical questions. These assume the form of, "If . . . (biased or ridiculous assumption) . . . would you maintain that . . . (charge or concern)?" For instance, "If septage had no microorganisms or chemicals, would you maintain it presents any hazard?" If the lawyer of the opposing side asks something like this, it is because he has in mind showing to the jury that in the case at hand septage has not been shown to have hazardous constituents and hence poses no hazard. One should not answer a hypothetical question with a brief "yes" or "no." Juries may be swayed by such an answer. A good tactic is to reply with a hypothetical answer in order to convey the proper perspective. For instance, one could reply, "If lead were lighter than water, it would float, and if septage had no hazardous constituents, I would not be concerned." Often, the lawyer on one's side will object to hypothetical questions before one answers them.

A witness must keep in mind that when the opposing lawyer is being very nice, that's probably to bring the witness's guard down. That same lawyer may try to plant an idea in the witness's mind and ask innocently, "Isn't it so?" And subconsciously the witness will tend to agree, as one is generally polite to a charming fellow. Of course, this can be deadly when a court reporter is taking notes. Later on the

---

*About 20 years ago I read in *Psychology Today* about an interesting experiment. Two speakers (actors) addressed an audience. The first one delivered a common speech. The second one's speech consisted of completely nonsensical sentences, on purpose, but he spoke with conviction and in an amiable tone. The audience felt that the second speech was better and held a far more favorable impression of the second speaker.

same nice lawyer may treat the agreeable witness like scum of the earth.

If and when the opposing-side lawyer bullies, insults, or denigrates you or your qualifications in order to make you angry so that you will not be able to think clearly, keep your cool by remembering that he is saying much about himself and nothing about you.

The meaning of any one specific word may not be the same to the parties in conflict. Once this is clear, some of the heat and conflict disappear. For instance, one party may maintain that sewage is beneficial; another one that it is a hazard. The former might think of sewage as "night soil," or fertilizer, the latter as a public health hazard. Try to visualize words as having subscripts, $sewage_1$, $sewage_2$, $hazard_1$, $hazard_2$, etc. and keep in mind that each may mean something different to someone else. Make sure that whoever cross examines you defines his words to your satisfaction.

A witness may be influenced by who is asking the questions (or by where the questioner is "coming from"). As a hypothetical example, let us assume that I saw two pink sparrows in a flock of 100. If anyone were to ask me, "Were there a lot of pink sparrows in the flock?" I would answer, "No." (Two out of 100 is not a lot.) If later on someone I knew was an ornithologist would ask me the same question, I would answer "Yes," truthfully, and without giving it any thought. That's because subconsciously I am aware that ornithologists know that the incidence of pink sparrows is, say, one in a million sparrows. (So two out of 100 is a lot.) Though truthful, my contradictory answers would have made me look like a liar in front of a jury. The way to avoid this type of predicament is to give factual or quantitative answers like, "I saw two out of 100."

And you should be aware that people who act as advocates (and take sides) have trouble thinking in objective ways that lead to problem resolution. As a theoretical illustration, let us examine the thinking of honorable people, some of whom are for and some against capital punishment as a deterrent to crime. A typical "either-or" situation: Each side sees itself in the right. One side may prove its point with statistics show-ing no decrease in crime right after a public hanging. The opposite side may point out that dead criminals are not recidivists. Both sides fail to notice that statistics are unreli-able unless they refer to subpopulations, e.g., the death pen-alty may stop some types of people from indulging in "white collar" crime, but it would do little or nothing to prevent crimes of passion. As in many other disputes, I see this as a problem of properly defining what's desired and determin-ing the social cost and benefits of applying a given reward, punishment, or other treatment to carefully defined subpop-ulations, so as to achieve desirable results at acceptable costs and hazards to society. In this frame of mind, it is easier to see that opposing sides may base themselves on accurate perceptions while using the wrong frame of reasoning. (Of course, regarding the death penalty, things are more compli-cated. But this is just an illustration.)

In general, lawyers of opposing sides will be quite civil to each other, in and out of court. But don't be misled: their aim is to win, and some will do just about anything to achieve this aim. Nowadays, there are companies that coach attor-neys and expert witnesses on defending against dirty tricks and on increasing the credibility of testimony. In the PBS TV broadcast "Nightly Business Report" (October 25, 1989), a representative from one of those companies demonstrated how to destroy an honest witness's credibility:

> Lawyer: Do you have kids?
> Witness: Yes.
> Lawyer: Did you tell them that there is a Santa Claus?
> Witness: Yes.
> Lawyer: I hope you won't lie now that you give testimony in court.

Finally, my advice to prospective expert witnesses is to remember that they are a tool in the hands of their attorney. (This view is not shared by Matson.[2] Perhaps his "tool$_1$" is not my "tool$_2$". It is advisable to read Matson's book to obtain a broader perspective.)

REFERENCES

1. Brown, S. M. 1983. The environmental health professional as an expert witness. *J. Environ. Health* 46:84–87.
2. Matson, J. V. 1990. *Effective Expert Witnessing*. Lewis Publishers, Inc., Chelsea, MI.

ARTICLE ABOUT EXPERT WITNESSES

By kind permission from the *Journal of Environmental Health*, Dr. S. M. Brown's article is reproduced here. It is relevant to expert witnesses in any field or specialty. The material below was based on information from a subsequently published textbook, *Environmental Health Law* (Brown, S. M., and T. R. Forrest, Praeger Publishers, New York, NY, 1984).

The Environmental Health Professional As an Expert Witness

Sanford M. Brown, Ph.D.

The sanitarian, environmental health specialist, as well as other health practitioners, may be called from time to time to serve as an expert witness. Traditionally, health professionals have been utilized by courts to help resolve cases involving scientific fact and technical issues where there is a difference of opinion on the "best available technology." This has been the case in the environmental health field, especially during the 1970's—the decade of the environment. Consequently, expert opinion plays an important role in the field of environmental health due to the inexactitude of current knowledge in the area and opens the door for the environmental health specialist to be utilized as an expert witness. Although sanitarians or other environmental health practitioners may be generalists, they usually specialize, over the years, in a particular area of environmental health field practice where their expertise can be utilized.

An expert witness is one who, because of education and experience, demonstrates a certain knowledge not possessed by the ordinary person. The Federal Rules of Evidence, Rule 702, state that an expert is a person who may give testimony in the "form of an opinion or otherwise," where he or she is qualified "by knowledge, skill, training or education," if "scientific, technical or other specialized knowledge will assist the trier of fact to understand the evidence to determine a fact in issue."[9] The expert assists the judge and/or jury to understand and interpret technical evidence in a specialized area. The expert differs from the lay witness in that in addition to the stating of fact, the expert is permitted to offer conclusions, draw cause-effect relationships and express opinions.[10] The expert may be used in discovery proceedings, the preparation and completion of interrogatories and/or depositions, and as a witness in court proceedings.

Table 1 summarizes the criteria utilized for the selection of an expert witness. Two basis criteria apply to the selection of experts: A) the expert must qualify as an expert in the *specific area* in which he or she will testify and, B) the expert must be able to explain simply his or her theories in a court appearance.[9] Generalist sanitarians, qualifying as expert witnesses, must have a specific area of expertise based on education, experience, knowledge, skill or training. Many times a sanitarian may have a limited speciality or be a program specialist of consultant in a specific area of environmental practice. These practitioners are best qualified as experts. Education and experience are the keys to expert participation although either may be the basis of qualification and not necessarily both.[3,9]

Table 2 lists the guidelines for the expert's preparation. The

Table P.1 Criteria for Selection of an Expert Witness[2,9]

1. The expert should qualify as an expert in the specific area in which he or she is to testify.
2. The expert should be qualified to appear in court to explain his or her theories and be cross-examined.
3. The expert should have a comprehensive understanding of his or her area but should also have the ability to explain technical findings in a simple, clear and concise manner.
4. The expert should be able to reduce his or her technical knowledge to understandable English.
5. The expert should have the ability to think on his or her feet.
6. The expert should have the ability to deal in facts and theories.
7. The expert should be a professional and a practitioner.
8. The expert should have both public and private sector experience.
9. The expert should be familiar with the literature in the area of expertise.
10. The expert should have published in the area of his or her expertise.
11. The expert should be a local or national expert, appropriate to the needs of the class.

attorney and the environmental health expert must undertake thorough pretrial preparation. They must work closely together on technical matters and corresponding legal issues. The attorney and the expert should review all field notes, tests, raw data, inspection forms, reports and correspondence generated as a result of the inspection reports, calculations, prior testimony, depositions, interrogatories, pertinent literature and publications. When the expert is engaged in civil litigation, much of the expert's time will be devoted to discovery proceedings. Discovery is the process by which each party to a case is allowed to learn of and probe the factual components of the other party's case. Generally, discovery allows one party to examine the other's witnesses and documents, etc. before the trial, either orally or by written interrogatories. Many times the expert undergoes the preparation of a deposition, which is a written record of sworn testimony, made before a public officer for purposes of court action. The deposition is usually in the form of interrogatories posed by an attorney.

Table P.2 Preparation of Experts—Pre-Trial Preparation[5,6,9]

1. Before you testify, visit a court and listen to other experts testify.
2. Do not participate in a case against your best judgment.
3. Know the client before agreeing to work for him or her.
4. Know the attorney before agreeing to work for him or her—do not risk your reputation with a careless attorney.
5. Prepare yourself thoroughly in advance with field and/or record investigations.
6. Document conditions thoroughly with drawings, maps and photographs that demonstrate facts, interpretations and conclusions.
7. Review your testimony with the attorney well in advance of the case.
8. Educate the attorney thoroughly about the technical facts of the case.
9. Plan the presentation of your testimony with the attorney in advance.
10. *Be prepared*. Review all works and reports on the issue as well as your professional writing that may be related. As a rule, allow 5 to 10 hours of participation for each hour on the stand.
11. Follow the directions of the attorney.
12. Coordinate your testimony with other experts on your team.
13. Review pretrial depositions of witnesses and experts in your area of testimony.
14. If the opposing attorney takes your deposition as an expert, be prepared to supply your notes, correspondence and reports that you used as a basis for your information.
15. Review transcripts of relevant court proceedings prior to your appearance in court.
16. Do not memorize what you are going to say.
17. Do not stray beyond your area of competence.
18. Testify for the benefit of lay triers of fact.
19. Do not have a disdain for laypersons or the adversary's witnesses or attorney.
20. Familiarize yourself and the attorney with the literature in the area.

The pretrial preparation must include the basis for the expert's opinions. The expert must be prepared to give opinion on facts in evidence and on facts not in evidence but reasonably relied on by experts.[9] The attorney should review with the sanitarian the following traditional rules of expert testimony.[11]

1. Expert testimony on the matter directly at issue is inadmissible, particularly if the issue is a mixture of fact and law.[12]
2. Expert testimony is not admissible if it deals with matters of common knowledge.[8]
3. Facts upon which an opinion is based must be established by evidence.[7]
4. The adverse party's expert cannot generally be called for direct testimony, for an expert cannot be compelled, against his will, to render expert testimony.[4]

Expert testimony begins with the traditional qualification of the witness to stipulate the witnesses' qualifications and concede expert status. This concession should be resisted when the qualifications of the environmental health specialist are impressive. This is no time for modesty. Impressive credentials should be entered into the court proceedings. The witness is usually asked a series of qualifying questions to establish his or her expertise. The questions may include the following:

1. What is your name? Where do you live?
2. What is your profession? What does the profession involve? Do you specialize within your profession? What does the specialty involve? How long have you practiced your profession?
3. With respect to your formal education, what colleges and/or universities did you attend and what degrees did you receive? Were those degrees in any specialized area? If so, please explain? Was the program accredited? What other schools have similar programs in this area? Are they accredited programs?
4. With whom are you employed? How long have you been employed in your present position? What are the duties of your position?
5. Are you registered, certified, licensed, etc.? In this state? Are you licensed, etc. in any other state? How long have you been licensed? Have you been in practice all that time? Are you certified in a speciality? If so, what does that involve?
6. In the course of your professional duties, have you per-

formed analyses or tests? What did this involve? How
many tests have you conducted?

7. Have you done any teaching? How long? Describe your
teaching responsibilities?

8. Have you published any written works in the area of
your expertise? Have these works appeared in refereed,
professional journals? Would you state the title of some
of these works? Do you have an up-to-date and accurate
list of your publications? Have you reviewed written
works for a professional journal?

9. Are you a member of any professional associations? Do
you hold any offices in those associations?

10. Have you performed consulting duties in your area of
expertise? For the government? For private companies?

11. Have you received any prizes or awards in your field?
Please describe a few of those awards.

12. Have you ever previously testified as an expert witness in
an administrative or judicial proceeding? On how many
occasions?

After the questioning, the attorney may offer the witness
as an expert. After the offer, the opposing party may challenge the qualifications of the expert and ask questions concerning the qualifications. The court then makes the decision
on the expert witness.

Once qualified, the expert witness may testify to facts perceived by the witnesses and then proceed with opinion testimony. Table 3 lists the guidelines for the appearance in court
and direct testimony of an expert witness. The expert then
gives his or her opinion or sometimes a description of certain
conditions. The expert's attorney may utilize the hypothetical question procedure in order to more fully illustrate to the
judge and/or jury the facts the expert is relying upon. This
procedure includes a series of questions that are based on
and begin with a series of assumed facts. The facts are listed
in a specific order and the expert witness is questioned on
the facts.

The presentation of direct expert testimony usually follows
one or two methods: A) testify to all evidence bearing on the

Table P.3 Court Appearance—Testimony[5,6,9]

1. Stand upright when taking the oath. Say "I do" clearly.
2. Use notes to refresh your memory on lengthy data, and that of others, but do *not* use an outline of your testimony. The court may ask to see the notes.
3. When answering questions, look at the attorney and listen carefully to the question; then, answer to the judge and/or the jury, not the attorney.
4. Convey the image of competence and sincerity.
5. State your professional qualifications *clearly* and *completely*. Include education, experience, professional membership and professional registration or license.
6. Display total impartiality, don't be a surrogate advocate.
7. Speak in clear, loud tones in an authoritative manner.
8. Use plain language. Omit uncommon or technical terms or at least use an illustration that will clearly convey your message.
9. Answer only the question asked and don't expand on your answer.
10. Don't be content with a brief answer to a complex question, clarify the basic points and conclusions to the question.
11. Help your attorney rephrase a question that is not clear. You can say "Do you mean . . .?"
12. Do not exaggerate your response.
13. Act modest on the stand.
14. Do not guess; if you do not know the answer, say so.
15. Do not be afraid to admit a mistake or qualify an answer. A reputation of honesty and sincerity is highly valuable.
16. If you answer "sometimes," or "usually not" or "under certain circumstances," you may indicate to your attorney to elaborate on the point in question.
17. Be serious and polite. Never lose your temper.
18. Don't say, "That is all I remember." You may think of more later and then you can add it.
19. Don't surprise your own attorney by your testimony.

case during direct examination by your attorney, or B) delay introduction of some of the evidence which is damaging to your adversary in anticipation that the opposing attorney will request it during cross-examination, in belief that you omitted a statement on the subject because it contradicts your conclusions. If the opposing attorney does not ask about this information, your attorney should question you on this information during redirect examination. Following

Table P.4 Court Appearance—Cross Examination[5,6,9]

1. Do not rush your answers. Never allow the opposing attorney to force you to rapid conclusions, and allow time for your attorney to object.
2. Do not accept confusing rapid-fire questions. Be deliberate and ask the attorney to repeat the question.
3. Do not attempt to answer several questions at once; have the attorney choose one to answer.
4. Beware of trick questions.
5. Do not hesitate if the answer is obvious.
6. Do not allow the opposing attorney to disturb your composure.
7. Do not try to be clever.
8. Do not forget the opposing attorney may be better informed on some point than you are.
9. Watch out for certain ''stock questions.'' ''Did your attorney tell you to say that?''
10. Do not be too eager to agree with authorities in your field.
11. Limit your answers to the questions asked.
12. Do not risk confusion with a long, hypothetical question, ask the court reporter to repeat the question.
13. Do not be misled by compound questions.
14. Remain cool, objective and as cooperative as possible. Many judges don't like uncooperative hostile witnesses.
15. Judges themselves may inject their own questions, and the witness should be fully cooperative and duly deferential.

the direct testimony, the attorney and the expert should quickly review the strength as well as the weaknesses of the expert's testimony in preparation for cross-examination.

During cross-examination, the opposing attorney will usually treat an expert witness in one of three ways: A) as if the expert is well-prepared, knowledgeable and truly an expert; in this case, the attorney will ask few questions because he or she fears damaging answers, B) as if the expert does not know his or her subject or the facts of the case; in this case, the attorney attempts to discredit the witness, C) as if the expert is unsure about the important facts or aspects of the case; in this case, the attorney attempts to discredit the testimony and obtains conflicting or contrary statements for the

Table P.5 General Information—In Court[5,6,9]

1. Do not appear in court until you are instructed to do so.
2. On the witness stand, sit erect and be alert at all times.
3. The court will look at your appearance: experts command high fees—dress accordingly.
4. Wear clean clothes and dress conservatively.
5. Follow the rules of the court even though they may seem restraining and inexplicable.
6. Do not discuss the case in the halls of courtroom.
7. Do not discuss or consult with opposing experts about the case.
8. Do not sit at the counsel's table inside the railing unless asked to do so.
9. Unless requested by your attorney to advise him or her during the trial, leave the court when you have completed your testimony.
10. Do not engage in lengthy note-passing with your attorney. Wait until the next recess to relay information.
11. Never try to win the game of "wits" with the opposing attorney. A good attorney never asks a question unless he knows what the answer should be.
12. Expect delays—don't get frustrated with them.
13. Never volunteer any public comment concerning the case to press, radio, T.V., etc.

record. There is no doubt that the environmental health expert must strive to represent the first category.

Table 4 summarizes the guidelines for the court appearance during cross-examination. During the opposing attorney's cross-examination, the expert witness should utilize his or her responses to add to his or her own direct testimony. Watch for the opposing attorney's effort to show your bias or improper interest in the case. Don't let the attorney establish control over you by asking questions that only require a yes or no answer; elaborate where necessary. Don't get caught in the lack-of-firsthand-knowledge trap. If at all possible, go and obtain firsthand knowledge.

Table 5 summarizes general information that the expert witness should follow in and around the court room. The sanitarian must remember that expert testimony is a team effort. You should only appear and testify when asked or told to do so. Remember that you are participating in an area where you are a novice. You must have the knowledge and also the ability to follow the rules of the court. Expert testimony in environmental health cases requires thorough preparation and full participation to be successful. It is an experience that a qualified sanitarian will never forget.

REFERENCES

1. Baldwin, M., and J. K. Page. 1970. *Law and the Environment*. Walker and Co., New York, p. 432.
2. Beck, D. 1977. The role of the expert witness in the environmental litigation. *Litigation, the Journal of the Section of the American Bar Association*. 3:38–40.
3. Bockrath, J. T. (1977) *Environmental Law for Engineers, Scientists and Managers*. McGraw-Hill Co. New York, p. 359.
4. Cold Metal Press Co. V. U.S. Eng. & Foundry Co. 83 F. Supp. 914 W. D. PA (1938). In Baldwin, M., and J. K. Page 1970. *Law and the Environment*. Walker and Co., New York.
5. Ember, L. R. 1975. The expert witness. Env. Sci. and Technol. 9:620–621.
6. Fahye, R. P. 1975. A guide to being a witness. *Ohio J. Environ. Health*. 15:25–26.
7. Mozer V. Aetna Life Insurance Co., 126 F 2nd 141 3rd Circuit (1942). In Baldwin, M., and J. K. Page. 1970. *Law and the Environment*. Walker and Co., New York.
8. Noah V. Bowery Savings Bank, 225 NY 284 (1919). In Baldwin, M. and J. K. Page. 1970. *Law and the Environment*. Walker and Co., New York.
9. Riesel, D. 1982. Pre-trial discovery of experts, expert testimony and examination of experts in environmental litigation. *Environmental Law*. Environmental Law Institute, Philadelphia, p. 444.
10. Sikora, V. 1981. Proving environmental health violations. *J. Environ. Health*. 43:195–200.

11. Sive, D. 1970. Securing, examining and cross-examining expert witness in environmental cases. *Michigan Law Review*. 68:1175–1198.
12. U.S. V. Roberts, 192 F. 2nd 893 5th Circuit (1951).

APPENDIX Q

CASE HISTORIES OF LEGAL CONFLICT

Sanitarians, geologists, engineers, lawyers, and other professionals may become involved in lawsuits regarding septic systems. This appendix illustrates such involvement with real case histories, given in detail sufficient for readers to rely on their own insights and form their own opinions

In order to preserve anonymity, the gender of some of the protagonists has been changed, and the protagonists are represented by the following abbreviations:

- P = plaintiff(s)
- PL = plaintiff's lawyer(s)
- PLS = plaintiff's lawyer(s) representing a state (where the state is a plaintiff)
- D = defendant(s)
- DL = defendant's lawyer(s)
- CSSS = county septic system specialist
- SSC = septic system consultant (private)
- EW = expert witness
- G = geologist
- E = engineer
- EG = engineering geologist

Q.1 DISCIPLINARY ACTION AGAINST PROFESSIONALS

Q.1.1 An Engineer

A group of neighboring counties had formed a committee composed of local consulting engineers that met monthly with county officials to standardize percolation test requirements. At a meeting of this committee, CSSS saw a new member in attendance, E. One of those present addressed E and accused her passionately of being the "disgrace of the engineering profession." The other committee members, (very) civil engineers, tried to calm things down. (Much later, some of them told CSSS that they believed E routinely manufactured reports with fake data, precooked at her office.) E didn't come to any more committee meetings.

A year or two later, state consumer affairs investigators contacted CSSS and asked him to cooperate in their investigation of E. CSSS agreed and provided them with reports submitted by E. (CSSS had rejected these reports because in his opinion they contained false data.)

CSSS received a list of charges against E and a notice to appear at the disciplinary hearing (in front of an administrative law judge). CSSS read the charges. Two of them related to not reporting shallow groundwater (at a depth of about three feet) at two sites that supposedly had had exploratory excavations down to at least 8 feet below ground. Another charge related to exploring and allegedly approving a site for a tract subdivision; the houses were falling apart, the foundations were cracking and sinking, and tires and trees found below foundations identified the site as a former landfill. Other charges were based on the reports CSSS gave to the investigators.

CSSS went to the disciplinary hearing. There was no jury. E did not bring an attorney. She plea-bargained with the judge, stating that she would admit guilt regarding the charges of not reporting shallow groundwater. But she would not address anything relating to the landfill affair on the grounds that she was the subject of another related lawsuit on this matter, and anything she said could be used to incriminate her.* And she refused to address any of the charges based on the reports CSSS rejected, on the grounds that a sanitarian (CSSS) is not qualified to pass judgment on the work of a professional engineer. And with a smile she showed the judge a beautiful commendation certificate given her by CSSS' county. (She had left her name, address, and signature on the percolation test standards committee's roster on the occasion she was in attendance; the county mailed commendation certificates to every person named on that roster.)

The judge accepted her plea. E's license was suspended for six months. The local press reported contacting developers who claimed E's firm guaranteed producing favorable reports, even before performing field explorations.

CSSS learned that, if his county ever had to press charges against a professional, it would be advantageous to enlist the services of a member of the same profession as the accused. (And also that commendation certificates should be given sparingly and only for good cause.)

Q.1.2 A Geologist

An engineer, E, wanted to do perk testing. He assured CSSS that he was an expert.

The reports E began to submit to CSSS clearly indicated that E was not competent. E's reports failed to meet minimum code requirements. In his last submittal, E's leachline design was oblivious of steep slopes at the site of installation: the leachlines would have been installed either at exces-

---

*In 1988 an engineering geologist who investigated why the homes on the tract in the landfill were breaking up told CSSS that the other lawsuit never took place, as there was no way of recovering the cost of trial plus roughly $4 million in damages from E; a financial institution absorbed all the loss. E's company is no longer listed in the telephone directories. Allegedly, she moved out of the area.

sive depth or partly above ground! CSSS employed the usually effective technique to get rid of incompetents: CSSS told E that if he insisted on submitting substandard perk reports, CSSS would file a complaint with the professional registration authorities. CSSS wrote him a letter to this effect. (*The significance of this letter will become clear later on*.) E replied that he would no longer submit perk reports and that he would bring to the area a knowledgeable engineering geologist able to prepare such reports. And that's what he did.

The engineering geologist, let's call him "G," turned out to be a very charming fellow. He told CSSS he had a master's degree, and impressed him with his knowledge and competence. His reports seemed very good. G and CSSS became rather friendly.

A couple of years later, after a tip from a local building inspector, CSSS found that three houses (that were to be built on three small contiguous lots) had leachfields in excessively steep terrain. Their perk report had been prepared by G. G had seriously underreported the steepness of the slopes. Two young fellows who were building three "spec" houses (speculation houses for sale) on the lots asked CSSS to help them out, as they could not afford a lawsuit against G to recover their losses. CSSS helped them to salvage two of the three lots. The third one, with a 60% slope (reported as 40%) was unsalvageable.

CSSS went to the field and checked recently-approved reports submitted by G. CSSS found a pattern of underreporting of steep slopes. And, when he checked a small lot for which G had written a favorable report, CSSS found that a 55% slope (unsuitable) was reported as much less steep. And, even worse, an igneous rock outcrop that occupied much of the lot had been reported as "scattered boulders." Leachlines can't be installed in unweathered, solid igneous rock. CSSS asked EG, a county engineering geologist, to read the reports and to check all the sites. EG did, and took pictures. CSSS asked EG his opinion of G based on the perk reports G submitted, and EG answered that he believed the errors in G's reports were intentional and not due to incompetence. Also, he identified the rock outcrop as the top part of a dike (a massive, mostly buried "wall" of igneous rock).

CSSS wrote G a letter "inviting" him to appear at an office hearing and show cause why the county should keep on accepting his reports. CSSS thought that G would admit the obvious, that he would quit submitting reports, and that he would not repeat his mistake in other jurisdictions.

Instead, G hired an attorney to fight the charges at the office hearing. This made CSSS think that G was neither repentant nor concerned about ethics. CSSS no longer tried to save him from himself, and filed a complaint with the professional registration board.

At the office hearing, G's attorney tried to discredit CSSS's motivations and field observations. According to this attorney, CSSS was motivated by a personal desire to harm his client.* As for CSSS's field slope measurements, the attorney said that these were merely "CSSS's opinions."CSSS interjected, "Measurements are not opinions."**

The county geologist, EG, testified as to what he saw and measured in the field and showed the pictures he took at the sites. He corroborated CSSS's observations. The hearing officer was persuaded. County authorities notified G in writing that he was barred from submitting perk reports to the county.

G moved to a neighboring county. His charming personality must have worked for him. He became the chief geologist of a reputable engineering firm. The head of this firm phoned CSSS and protested politely CSSS's refusal to allow his highly qualified chief geologist to submit reports. He did not believe CSSS's explanations that he had good reasons to refuse reports signed by G.

---

*It is always so, at every hearing: The client is always innocent, and the county and especially CSSS has something personal against the client.
**One might get different opinions if one gives a glass of water to two people and asks, "Is the water warm or cold?" But there is no room for subjective opinions when someone sticks a calibrated thermometer into the water and replies, "It is 78°F." This is one reason why the scientific approach yields more truthful results than zealous advocacy.

About a year later, CSSS and EG received subpoenas to appear at G's disciplinary hearing on the basis of a complaint filed by the geologists' registration authority with the Attorney General's office. CSSS tried to find EG's pictures and the tape recording of the office hearing proceedings. They were absent from the storage file, and no one knew their whereabouts. CSSS prepared himself for the hearing the best he could.

Just before the hearing, EG and CSSS met with PLS, a state deputy Attorney General who had just flown in from the state capital. PLS told them about her immense work load and said that she could devote only two days to this case, including travel time. She reviewed the materials CSSS had gathered and EG's notes. They entered the hearing room and saw the defense team: G, G's new lawyer DL, and E!

CSSS put two and two together and concluded that the state's probability of success in its case against G was very low:

- CSSS did not have the pictures shown at (and the tape recording of) the office hearing.
- DL had all the time in the world to prepare plausible defenses, while PLS's time was extremely limited.
- CSSS had heard that no geologist had ever been disciplined before in a court of that state.

The proceedings began in front of an administrative law judge. There was no jury. CSSS was cross-examined about his qualifications and the evidence he produced. Then G testified. G and DL tried to convince the judge that CSSS had a personal grudge against G. According to G, the rock outcrop was just loose boulders that had rolled down to the site from uphill; slopes could be measured in ways that yielded different results, and anyhow there was no requirement for accuracy. CSSS testified that if boulders had rolled from above, the boulders would have empty spaces and/or soil between them; CSSS had seen only fractures in the outcrop, and the fractures contained no soil. And CSSS said that G's underreporting of slope steepness could not be anything but intentional. For instance, when the slope was mild, 20%, it was reported as 20%. But when it was steep, say, 60%, it was reported as 40%, and a 40% slope was reported as 20%. CSSS said that anyone could get better results with a carpenter's level, and pointed out that the county testing procedures handbook had a table that required determination of slopes to within 5 percentage points.

G again testified as to his considerable expertise and dismissed CSSS's allegations that G had harmed the spec house developers: "Nobody has sued me," he said. Then E testified (for G). He described himself as a very experienced engineer. He stated that slope readings vary according to how one holds the level. The judge asked him, "Do you mean that the way you hold the level results in different slope readings?" "Yes," replied E with a smile. The "expert" had spoken; the judge must have accepted E's explanation, because from that moment on the judge addressed CSSS in a harsh tone of voice. CSSS felt like yelling "Hey, judge, if you pay for six pounds of groceries and you get four, would you accept the explanation that what you get depends on how the grocer holds the scale?" Of course, CSSS kept his mouth shut. The day was over, and the hearing was continued to a date months into the future. CSSS told PLS what CSSS knew about E, and later on CSSS mailed PLS a copy of the menacing letter he had sent to E years before.

The day of the second hearing CSSS entered the hearing room with EG and PLS, resigned to G's victory. And then he noticed that the pictures he thought had been "lost" were in the hands of G's team. And CSSS also noticed that E had a topographic map that E prepared for one of the lots in dispute. CSSS examined it. It corroborated CSSS's slope measurements, not G's! CSSS discussed these findings with PLS and EG.

PLS cross-examined E in front of the judge. She asked E, "Do you remember when you quit doing percolation testing in the county?" He did. "Do you remember why you quit doing percolation tests in the county?" E replied that he just wasn't interested, it was too much of a hassle, so he quit. PLS gave him the copy of the letter CSSS had sent E. "Please read it. Does this refresh your memory as to why you quit?" The effect was devastating. E replied in a loud and shrill voice trembling with alarm, "Where did you get this? Where did you get this?" PLS was firm. "Read it," she ordered. E read it. And in so doing, he revealed that he could have harmed his clients if CSSS had not detected his errors, and also revealed CSSS's threat of reporting him to the state registration authorities. PLS asked, "Who signed it?" "CSSS," replied E. "Does this refresh your memory as to why you quit doing percolation reports?" With a panicky voice E sputtered that his clients were not harmed, etc. As he rambled on and on and on, his credibility evaporated. After this event, the judge did not address CSSS in a harsh tone of voice.

EG took the stand. He stated his qualifications as an engineering geologist with a master's degree. He gave superbly professional testimony and corroborated CSSS's findings. PLS introduced as evidence the pictures in possession of G's team and asked EG to explain why he took each picture, which he did. (PLS said that these pictures were more credible now, as they were obtained from the defense.) EG stated that in college, sophomore geologists must pass a course on the use of the Brunton compass, which is used (as a level) to measure slopes. PLS borrowed from G's team the topographic map prepared by E and gave it to EG. "What slope did CSSS measure at such-and-such point on the lot?" EG checked and replied, "40%." "What slope did you measure?" EG checked and replied "40%." "What's the slope measured by E?" EG checked the map and replied, "40%." "What slope did G report at the same location?" EG checked and replied, "20%." Finally realizing the extent of the deception, the judge said, "Uh, that's enough." The judge asked both DL and PLS to figure out what punishment had been given in similar cases, and to report to her.

As PLS, EG, and CSSS left the hearing room, they overheard someone in G's group say, "We should not have brought the pictures." That's the only sign of repentance CSSS detected.*

What CSSS learned from this case is that:

1. In the past CSSS had had occasion to be sorry when he was too merciful to a friendly yet dishonest professional. This time CSSS felt that he did the right thing. CSSS felt that one should not give another chance to unethical people. It's unlikely they'll become ethical after a mere reprimand. They may try to come back with legal ammunition and fight the merciful (county official and) county; and then the county may find itself in a difficult legal position for not having clamped down hard when it should have. CSSS felt that unethical professionals should be given enough rope to hang themselves and then be reported to the pertinent registration board with no qualms. The board's own disciplinary system probably acts with more than enough leniency and mercy.

2. Evidence should be guarded with someone's life, at least in triplicate, and in at least two different secret places.

3. Justice might not have prevailed if it were not for luck, a procedures manual with well-defined requirements, the additional testimony of a well-qualified professional of the same profession as the accused, and the dedication and competence of an overworked deputy attorney general (may her free time and salary increase in proportion to the moral debt owed her by citizens of her state).

---

*About a year later CSSS found out that G had been subject to disciplinary action for negligence in misrepresenting certain material information to a local public agency. G's license was suspended for two months, but this suspension was stayed and G was placed on probation for two years under the following terms and conditions: (1) G must take courses related to percolation testing; (2) G can't personally measure slopes during the probation period; (3) G shall obey all federal, state, and local laws; and (4) time of practice or residency outside the state will not be credited. Also, G's engineering geologist certificate was "publicly reproved."

Q.2 A NEGLIGENCE LAWSUIT THIS SIDE OF HELL

Years ago, a private engineer told me that he was going to placate a client dissatisfied with a septic system he had designed. Since the client's complaint was totally absurd, I asked the engineer why he bothered: "Let him sue, he'll lose." The engineer replied, "No, when you are sued you lose, even if you win."

He knew what he was talking about. The story in this section illustrates why the prospect of being sued, even over unreasonable or false charges, can scare someone. It also offers an opportunity to describe how a lawsuit works against a professional. The actual events have been slightly modified; for instance, "PL" was really three different lawyers for the plaintiff. (Readers who are not familiar with the technical subjects discussed might suffer some boredom. Readers who are familiar might experience a temporary elevation of blood pressure. Watch out!)

A lawyer retained SSC as an expert witness. The lawyer, let's call him defense lawyer or DL, asked SSC to evaluate a perk report prepared by DL's client, a large, reputable *engi*neering and geological consulting firm; let's call this firm "Reng," as per the previous italic "r" and "eng." Reng was being sued by a developer on account of the alleged shortcomings of the perk report prepared by Reng (which proposed using multiple seepage pits) and the alleged damages these shortcomings caused.

After SSC read the perk report he met with DL. DL asked how SSC would have rated the report at the time it was prepared. SSC replied that given state-of-the-art knowledge at that time, it was definitely better than the average. (And that he would have given it a "B" grade, because he would not give an "A" to any report that based itself on the Uniform Plumbing Code though this is a common legal requirement.) Reng had indeed followed all the prescriptions of the then-current edition of the Uniform Plumbing Code. DL said that he was relieved to hear that the report did abide by this code. He asked SSC to prepare for a deposition.

A month or so later SSC entered the room where the plaintiff's lawyer, PL, was to take his deposition. There SSC also met DLA, a defense lawyer for another large and well-established engineering firm which was being sued together with Reng by PL's client. The charge against this other engineering firm was that it had used Reng's perk report to propose a more comprehensive septic system design, and had failed to detect the alleged shortcomings in Reng's report.

PL took SSC's deposition. PL appeared to be unfamiliar with septic systems. Many of the questions he asked were so nonsensical that some of SSC's answers were bound to be ridiculous, like baby talk. PL became upset with DL because SSC had not brought with him a report with "canned statements" revealing SSC's opinions. Apparently on purpose, DL had not pressed SSC to write a report; he just made SSC available to PL so PL would extract whatever information he could out of SSC, as DL had done himself.

In brief, SSC commended the report for revealing problems with the site, unlike a previous report (by another consultant) that revealed no problems at all.

The deposition went on from 10 a.m. to late in the afternoon. Instead of coming back another day so that PL could finish asking all the questions he wanted, the lawyers and SSC agreed to keep on going with the deposition. It ended late at night. DL was satisfied and said gleefully that after this deposition PL could not call SSC back and make further inquiries.

DL read *Septic Systems Handbook* and asked SSC to be present when he took the deposition of EW, PL's expert witness.

Weeks later, DL took EW's deposition in the presence of PL, DLA, and DLA's expert witness. The pay of expert witnesses and lawyers plus one court reporter added up to about $8,000–10,000 per each of the four days of deposition. And the lawyers had already deposed other people involved in this case. SSC saw DLA carrying his paperwork for this case on a dolly, four or five cardboard boxes stacked one on top of another, in a pile five feet high!

DL told SSC that he estimated pre-trial costs of all parties at $125,000 and court trial costs at $100,000 per each party. (These figures did not take into account the value of the time spent by Reng's personnel.) He also told SSC that Reng was being sued for well over a million dollars, and that the insurance company that insured Reng would not renew Reng's policy, even if Reng were cleared in a trial. Reng had to get another insurer and possibly pay much higher insurance fees, even if not guilty of anything.

The first day of deposition, EW described himself as an engineering geologist and hydrogeologist who was a private consultant to various engineering and geology firms. His consultations extended to the field of sewage disposal. After he said repeatedly and in many ways that each county had its own different Uniform Plumbing Code (there is but one for the whole world), SSC suspected that EW was anything but an expert. At a "bathroom-visit break," SSC told this to DL and suggested that he ask EW how many perk reports he had ever prepared. Later, EW answered that he had prepared one or two perk reports for leachlines, none for seepage pits.

EW produced copies of EW's letters to PL, in which EW pointed out the alleged deficiencies of Reng's report and blamed the other engineering company that used this report for not detecting the alleged deficiencies. In SSC's opinion, EW's allegations were absolutely preposterous.

For example, on the second day of his deposition, EW produced a bunch of photocopied pages. He had just read about seepage pits in a photocopied chapter from *Septic Systems Handbook* (given to him by one of his engineering acquaintances). He stated that reading this chapter confirmed his beliefs. However, it was obvious that he had not bothered to read the rest of this *Handbook*, because his statements were less than fully informed. For instance, he said that after seepage pits get clogged by biomat they are scarified with a steel brush, and that he read about it in the

Handbook! The documents EW produced included a schematic of a seepage pit cross section. If one didn't know anything about seepage pits, after looking at this schematic one would realize that seepage pits cannot be scarified.* At breaks, with the consent of DL, SSC shared impressions and opinions with DLA and his expert. This expert, a veteran civil engineer, had no trouble detecting on his own each and every error in EW's statements. By the end of the second day of depositions DLA told SSC that EW was "dead."

On the third day of EW's depositions, DLA did much of the questioning. To the chagrin of DL, DLA confronted EW with some of EW's previous ill-informed statements. (DL preferred to wait and do this at trial, so that EW and PL would not realize EW's goofs and have time to prepare plausible explanations .) EW admitted that he had never used the Uniform Plumbing Code until he was hired by PL in this case. And he also admitted that he did not know how to design septic systems. But he felt competent to declare that Reng's report and DLA's client's proposal did not meet the code. He had *never seen* something like Reng proposed, i.e., a septic system with multiple seepage pits for a cluster of houses. And he thought that this system could not work!

In SSC's opinion, EW appeared to be a fine gentleman who was not able to even suspect the extent of his ignorance about septic systems.

SSC asked DL if by now PL's client would realize that there were no grounds for a lawsuit, apologize, and withdraw the lawsuit. DL replied that if any of that occurred this late in the

---

*EW was familiar with sewage disposal ponds or percolation ponds. When the bottoms of these shallow open-air ponds plug up with biomat or with silt or clay, they are drained and allowed to dry; then they are raked (scarified) to break up and open the crust and restore infiltration. To do this to a seepage pit would require removing the inner concrete-block wall, using potent magic to prevent collapse of the structure (unsupported gravel and wet soil around cavity would collapse), removing the gravel envelope, scarifying the sidewalls who knows how, and reinstalling the gravel and inner wall. Even if this were feasible, it would be far cheaper to build a new seepage pit.

game, PL's own client could sue PL for malpractice for not realizing this fact earlier. And he said that EW could not be sued for the damage he had done, and that Reng could not recover damages even from PL. He suspected PL might still want to go to trial because juries are never completely predictable, and though the probability of PL winning was low, the "damages" prize claimed, now bargained down to $1.5 million, was worth a try. Now was the time when all attorneys try to reach a settlement satisfactory to their clients. He, DL, would contact SSC if SSC's services would be needed at trial.

Many days later, DL asked SSC to prepare for trial and informed him as to what had happened. DLA had settled separately and agreed to pay $25,000 to remove his client from the lawsuit.* The plaintiffs were willing to lower their demand to $500,000 (the amount covered by Reng's insurance). But Reng's principal was a man of principles. He was certain Reng was innocent and would not agree to concede one cent.

SSC learned that this negligence lawsuit was connected to another lawsuit started by other parties and that considering the complaints, cross-complaints, and delays by the many parties involved, a court trial *might* not occur during his lifetime.

It is worth noting that Reng's insurance company first delayed and then quit paying the fees of DL's expert witnesses. DL told SSC that this insurance company could get sued for substantial damages if the case was lost due to unpaid (hence uncooperative) expert witnesses. DL's firm was large and reputable and paid the expert witnesses out of

---

*This meant that DLA's client yielded to what some might describe as extortion, but SSC understood why. Assuming that the chance of losing the case was only 1 in 20 (that is, 5%), 5% of $1.5 million constitutes a payoff or value loss of $0.05 \times 1.5 \times 10^6 = \$75,000$, plus trial costs. Sadly, our judicial system allows this form of extortion. SSC felt certain that if judge-appointed experts would have heard and seen half of what SSC did, the lawsuit would have ended at once.

its own pocket. (Who knows what might befall "little guys" who are defended by impecunious law firms.)

One thing SSC learned from this experience is that, metaphorically speaking, the adversary system may lead many into the temptation of confrontation, and may lead some (including innocent victims) into financial perdition. Trivial or unreasonable charges can result in a mind-boggling waste of time and money. One way of avoiding this might be arbitration, as suggested in Appendix N. Another possibility might exist if professional engineering or like associations offered free "screening" and advisory services to courts, or had arbitration boards, or boards that would "screen" for a fee the validity of litigants' claims before a lawsuit commences. Or, if judges were represented by their own expert witnesses.

Q.3 A LAWSUIT INVOLVING SEPTIC SYSTEM FAILURE

A lawyer told me that in his experience, when the issues were "black and white," it was better to go to court than to arbitration. He perceived arbitrators as operating on the basis that no one is 100% right, as seeing everything gray, and as too eager to compromise. He might be right; I don't know. The following "black and white" case illustrates the difficulties of being 100% right and proving this in court. Although I will still use the abbreviations PL and DL, these refer to completely different lawyers. In fact, "PL" was two lawyers for the plaintiffs, and "DL" was two lawyers for the defendant.

PL retained SSC as an expert witness the very day of the trial. PL assured SSC that everything would be over after "about two hours" in court. PL explained the lawsuit. Her clients bought a house from a real estate broker who had owned and lived in this house. This broker told them that a previous "problem" with the septic system had been corrected. Five days after her clients moved into their new home, the bathtub and toilet backed up and poured sewage on tiles and carpets. Thenceforth they had to severely limit their use of water for years, until the time of the trial. The ex-owner (broker and defendant) refused to admit any blame or to pay even part of the repair costs, so he was being sued by PL's clients. PL took SSC to inspect her clients' property. In the small front yard, SSC saw a riser with sewage near its top. Then PL and SSC rushed to court.

At court, PL asked the judge to hear the case at once, as it had been "trailing" and it had been given priority the previous week. DL asked for a postponement as his client was to go on a trip to another state the following day, but the judge refused. The judge admonished all parties to try to reach a settlement.

DL offered to settle and pay for a partial repair; PL's clients were incensed and refused this offer, which they would have accepted years earlier, before they decided to sue. There was no jury, and all the lawyers agreed that they did not even need a court reporter.

The trial began with the testimony of PL's other expert witness, STP, who had been a septic tank pumper and contractor for over 30 years. He gave an outstanding presentation. In brief, he said that when the tract was graded in 1977, the soil was built up (and compacted) 5 to 15 feet in some lots, and that contractors often installed leachlines on compacted fill because neither the project engineer nor county inspectors were keeping an eye on them. So, a substantial number of houses in this (and in neighboring) tracts had failing septic systems one year after the houses were sold. He had repaired many such systems by installing seepage pits, which penetrated and extended beyond the compacted soil.

The following day, STP was cross examined by DL and did very well. He was a practical man, yet everything he said was in agreement with septic systems theory and practice. SSC could not have done a better job.

PL tried to have SSC testify. DL objected because SSC was

brought in at the last minute and DL had not had an opportunity to take SSC's deposition. DL had not objected when
he thought SSC was just another septic tank pumper, but he
objected now that he found out that SSC had a Ph.D. The
judge agreed with DL. PL told SSC to stay around and be a
"rebuttal witness." This type of witness can be introduced at
any time and without prior deposition, but only to rebut
someone else's statements. Very cleverly, PL intended to ask
every potentially damaging question of the defense's expert
witness, and thus give SSC an opportunity to voice all of his
expert opinions by means of rebuttals.

The defense's expert witness, EW, took the stand. He was
a septic tank pumper and contractor. EW puffed himself up
to be a very expert expert. Almost everything he said was
contrary to septic systems theory and practice. For instance,
EW said that compaction does not affect soil permeability,
because the soil texture is the same before and after compaction! (EW must have gotten this idea from the Uniform
Plumbing Code: it assigns soil absorption rates on the basis
of soil texture only.) When PL confronted EW with the
record of frequent septic tank pumpings prior to the sale of
the house, EW said that the record didn't mean that the
system was failing. The fact that sewage backed up into the
house five days after the new owners moved into the house
meant to him that the new owners "were using too much
water." He said that he could squeeze a cheap additional
leachline in the front yard and that this would solve the
problem. He said that he sized that leachline on the basis of a
map of the area, drawn by the county, that specified 25
square feet of absorption area (per 100 gallons of septic tank
capacity). As he said, one gets a permit and installs the
leachline. Easy stuff; he had done it many times and had
much experience. (SSC knew otherwise; the 25 square feet
figure seemed too low. SSC thought that if EW had installed
undersized leachlines in the area, he might have cheated his
customers and created public health hazards, as the leachlines were bound to fail in a short time.)

EW's cross examination ended late in the afternoon. As he

had done before, the judge again ordered all the witnesses to come back the following day. Before SSC left, SSC told PL what he thought of EW's expertise.

EW's testimony upset SSC so much that he could not sleep that night. He felt that the judge was giving as much credence to EW as to STP.

Early in the morning of the following day, SSC went to the county office and obtained copies of the requirements for the tract where the failing septic system was located. The official records current at the time the system was installed showed that the requirement was 45, not 25, square feet (nowadays, much higher). When SSC arrived at court, he gave the copies of the records to PL.

Soon after SSC began the rebuttal testimony, PL asked SSC to state his opinion about EW's presentation. SSC replied: "Despicable. When he wasn't lying he was speaking untruths. I was so mad about this that I couldn't sleep last night. . . ." (Here DL laughed.) ". . . He could be subject to a class action lawsuit." DL attacked SSC vociferously for daring to give a legal opinion, but the judge defended SSC vigorously and took an interest in what SSC was saying.

SSC read his notes and criticized EW's testimony point by point. Some of the points SSC made were:

- Per official records, the required design rate was 45, not 25, square feet.
- Compaction of soil does decrease permeability; even if the soil had not been compacted, the traffic of heavy equipment smudges the top of each layer of soil as the ground is built up. A thin smudged layer prevents downward flow about as much as if the whole soil were compacted.
- Once a septic tank is pumped twice a year (this did happen before the house was sold to the plaintiffs), county officials in EW's county consider this a system failure. It was strange that EW, a pumper, didn't know that twice-a-year pumping defines system failure, since every pumper has to provide a report of pumpings to county officials so that they can identify areas with septic system problems per that twice-a-year-pumping definition.

(While SSC was testifying, the defendant was squirming in his seat and his face assumed all the colors of the rainbow; but his hired guns, DL, were unperturbed and almost jovial.)

SSC was halfway through his notes and criticisms. The judge asked him directly, "What's wrong with EW's way of repairing the leachlines?" SSC replied that there were three things wrong:

- EW was proposing to install a substandard length of leachline, 53 feet in length per his incorrect calculations, though the minimum requirement calculated on the basis of 45 square feet (per 100 gallons of a 1200-gallon septic tank) was 77 feet long.
- EW was proposing to install a system where it was obvious that the soil was having problems absorbing the sewage. The previous leachline was having problems, and he wanted to repeat the mistake of the original installers, though he had noticed that there was a problem.
- There was barely enough space in the yard for the existing system. It would be difficult to impossible to squeeze in an additional leachline 53 feet long, much less one 77 feet long.

The judge took notes and said that SSC's testimony was no longer required.

PL asked SSC a few more questions. SSC stated that he would use seepage pits to get through the compacted or smudged soil layers.

PL prevailed. The judge told the lawyers to conclude and present their views in writing, and said that he considered the defendant to be particularly responsible because he was a real estate broker and had a duty to protect his clients.

One has to wonder about all the inconvenience that the litigants suffered. The plaintiffs were worried because the following day they were to depart on a cruise and feared they might have to cancel their trip. And, just in each of the days SSC served, each of the litigating parties had probably spent as much in lawyer's fees as it would cost to install remedial seepage pits. (The prevailing party might get reimbursed for lawyer's fees, though.)

If SSC had been just another septic tank pumper, the judge might not have known whose testimony to believe.

Q.4 LAWSUIT NEAR-MISSES

In this section we'll take a look at typical problems with septic systems practice and practitioners. The technical matters presented here are intertwined with ethical and with legal (lawsuit) considerations. Appendix R addresses the subject of ethical issues.

Q.4.1 One Power Auger

Some prospective home builders were distressed because their lots didn't pass perk tests (for leachlines). (The perk times measured by their engineer were about 80 minutes per inch; the county required a maximum of 60 minutes per inch.) So they retained SSC's services.

The test holes used by the builders' engineer had been drilled with power augers of the helical or screw type, commonly employed in the area. The test holes SSC used were prepared by hand; all smeared and compacted sidewalls were removed from his test holes. He got perk times of about 8 minutes per inch.

SSC's happy clients were mad at the engineer who had "failed" their lots. They asked SSC whether they should sue the engineer.

SSC took into account that:

- The engineer was not a crook. If he were, he might have "passed" rather than "failed" the lots.
- Just about every septic system consultant in the area was using screw-type power augers and treating the soil like

dirt. The engineer was just one of many, and he did not know better.

So, SSC told everyone that if he were in their shoes, he wouldn't sue. And they didn't.

Q.4.2 Another Nasty Power Auger

A very reputable engineering firm had done a perk report for a very expensive lot in an exclusive area. It drilled exploratory and test holes down to 50 feet with a screw-type flight auger; and it tested the holes per the falling-head test for seepage pits. The results showed an absorption rate of water per extrapolated 24-hr day of about 1.5 gallons/square foot per day, and an equivalent presumed absorption of sewage one-fifth as large, 0.3 gallons/square foot/day. The latter figure should have been 1.1 or higher to have met county requirements.* The engineers concluded that there was no hope for seepage pits, and that the owner could try leach-lines. However, they claimed that "testing for leachlines at the subject site would be costly and most likely provide results similar to those for seepage pits."

Another consultant dealt with the desperate owner of the lot, and retained SSC to see if the lot could be saved.

SSC went to the field and checked backhoe-excavated trenches. As he suspected, the strata were different from those reported in the original study. When one logs screw-auger spoils (soil that comes out of the drilled hole as the auger is drilling), it is difficult if not impossible to know how much of the spoils reflects materials drilled by the tip of the auger way down the hole, and how much reflects materials from the hole sidewalls up higher. Also, the silty soil encountered was of a type easily compacted and smudged by power augers.

So SSC tried testing for leachlines. His hand-excavated

---

*The county manual of procedures still does not acknowledge that these rates are pure fiction.

test holes yielded an excellent perk time of 3 to 4 minutes per inch. The lot was saved.

The happy lot owner was furious with the engineering company: Should he sue? He felt cheated. He had paid the engineers far more than he had paid SSC, yet the engineers wrote a report that "failed" his lot, while SSC had done the opposite.

In SSC's view, the owner was cheated, but not by the engineers. They did what practically all other professional consultants would have done. They followed the manual of procedures to the slightest detail. (This manual endorsed the use of screw-type augers.) And while they were looking at the manual, they didn't look at what their equipment was doing to the soil. SSC sent word out that in his opinion the owner didn't have a case. The owner didn't sue.

Q.4.3 Ability of Strata to Transmit Percolates

After SSC determined the rates of sewage disposal for proposed institutional buildings on a large lot with thin soil (2 to 10 feet thick), SSC saw that leachlines would meet county requirements. But estimates of the Darcy permeability constant led SSC to believe that the large amount of sewage to be discharged might not move out of the leachline areas without surfacing at the top of the ground. County requirements (and the Uniform Plumbing Code requirements) did not recognize or acknowledge the fact that after sewage percolates away from a leachline it has to move somewhere.

SSC's client requested that he certify the lot as suitable for all the proposed buildings. SSC refused. So, the client gave the job to an engineer who, though conscientious, was not familiar with the pertinent topics. This engineer certified the suitability of the lot, and the county approved it. The engineer followed procedures, and should not be liable for the possible consequences.

It should be noted that the ability of the soil to transmit percolate away from the leachfield can be estimated by means of Darcy's formula. This should be widely known, but isn't. I never saw this factor taken into account in any of the 3000-plus perk reports I reviewed. One exception: When I was a county employee, I forced a perk consultant to measure the Darcy permeability constant and to show site suitability before I approved his perk report. However, after a short while I gave up trying to force him and the other recalcitrant consultants to do likewise. I suppose that they were all afraid of liability if they performed "unusual" determinations not specified in the county procedures manual. Now that this problem is public knowledge, jurisdictions must take the lead in solving it.

Appendix R

ETHICS IN GENERAL AND IN SEPTIC SYSTEMS PRACTICE

Much of individual (or societal) behavior can be explained on the basis of the individual (or societal) ethics or morality. Though synonyms, "ethics" and "morality" have distinct connotations.

Per Webster's dictionary, the word "mores" is defined as "folkways that are considered conducive to the welfare of society and so, through general observance, develop the force of law, often becoming part of the formal legal code." "Moral" is defined as "relating to, dealing with, or capable of making the distinction between, right and wrong in conduct." Moral "implies conformity with the generally accepted standards of goodness or rightness in conduct or character." Its synonym, "ethical," implies "conformity with an elaborated, ideal code of moral principles."

Morality changes with time and geographical location. A few centuries ago, it was perfectly moral to burn at the stake witches, broadminded people, and even people who looked or acted too differently.* Morality is absorbed with mother's milk. One learns rules as to what is right and what is wrong, and that's it. No need to think: the rules are given. Most moral rules tend to be ethical, but not all at all times and places.

Ethics is a discipline of philosophy. It uses reason to determine what's right and what's wrong. One's outlook colors one's perceptions, and vice versa. In my attempts to understand what is and isn't ethical, I have developed personal views that might be colored in certain respects. It seems only proper to start this appendix by exposing my personal views so that their possible biases can be detected.

R.1 ETHICS IN GENERAL: A PERSONAL PHILOSOPHY

Over the last 35 years, I learned two things in my readings about philosophy (from Socrates to Bertrand Russell):

1. Historically, each school of philosophy had a logical way of reasoning, but the *assumptions* underlying its reasoning were mostly flawed, as proven by succeeding schools of philosophy. This perception allows one to dare question "logical" systems of belief, like, for instance, our judicial system's components and procedures and the wobbly assumptions these are based on. Questioning basic assumptions leads almost invariably to surprising results.
2. The philosopher I. Kant had the key to ethics. The key was to ask this question: "If everybody did as I intend to do,

---

*When I visited England, a tour guide told me that in the 11th century, shipwrecked French sailors came to an English coastal village. They were darker in complexion and spoke a weird language that no one could understand. It was obvious that they were from the Devil, and so they were burned at the stake. (A moral rendition of hospitality, for that time and place.) Not long ago the news media reported that Iranian children were sent across enemy minefields, with a promise that they'd go straight to Paradise when blown up, same as the adult holy warriors. All this is perfectly moral to anyone who grew up at the proper time and place.

would the world be a better or a worse place?" I have been using this key ever since Kant's words reached my ears in the 12th grade. It works in most situations, but some situations have no keyholes, no optimal solutions.

Before analyzing actual uses of this key in septic systems practice, let us expand our minds with some philosophical thoughts, and let us try to detect biases in the following philosophical views that color my decisions when confronted with ethical problems.

R.1.1 Everything is Related

Ethics (or lack of it) is involved in every human activity. Everything is related to everything else, and not just in the environmental sense. Past relates to present, language to view of the world, architecture to art, art to free time, free time to economic status, ethics to economic status, etc.

Our standard of living is very dependent on economic transactions, and economics is very related to ethics. As one instance, we might recall that in the 1970s the American car makers behaved as if their (ethical) mission was making money for their stockholders. The Japanese car makers behaved as if it was making good cars. As a result, by the early 1980s the Japanese were making cars and money, while the Americans were making little of either. As another illustration, a government's ethical aim, "doing the most good to the greatest number of people" implies defining the worth of—and distributing—what is considered good, and hence necessitates economics. Even at the personal level, ethics and economics are also intertwined. For instance, punctuality is the ethical regard for the value of someone else's time.

Septic systems practice (or for that matter, any other type of practice) results in the allocation of economic resources to achieve certain ends. The resulting patterns of land use (and misuse), as well as the allocation of economic resources themselves, may affect the economic viability of communities and even the mental and physical health of their inhabitants. In a competitive world, all these factors may affect the economics of a whole nation. Everything is absolutely related to everything else, directly or indirectly. Even if there were no other reasons for it, just the economic impact of ethics would suffice to merit a chapter about ethics in a book about professional practice.

R.1.2 Assumptions Tend to Be the "Clay Feet" of Elaborate Beliefs

If we assume that A = 2 and B = 3, then we can construct all kinds of mathematical relationships. For instance, the statement $(A^B/A) = (B - A) + B$ makes perfect sense, given our assumption. We can proceed to develop the whole world of mathematics from our simple initial assumption that A = 2 and B = 3. But if A is not exactly 2, or if B is not exactly 3, all of our derivations and constructions and statements are untrue, nonsensical, and mere pretense.

In a democracy, the reason for the existence of legal codes is an ethical or moral one: the welfare of society. But this point is often lost in a maze of laws, regulations, and regulatory procedures. These laws and regulations are like very complex mathematical constructs based on the implicit assumption that A = 2 and B = 3. Unfortunately, the "rule of law" has a tendency to become the rule of rules (i.e., the rule of the letter of the law), rather than the rule of reason (i.e., the spirit of the law, evident only if and when the assumptions that served as foundations for a law are acknowledged, like by checking whether A = 2 and B = 3).

I recall having a conversation with an attorney about the judge who threw out the case against an illegal drug manufacturer because of the "m" in the word methamphetamine was missing in the law as written (see Section 4 in Appendix M). This attorney said that the law had to be exact and precise, because otherwise judges could become too arbitrary. He explained that, if the common word "the" had been misspelled in a law, everyone would have understood what the

law meant; but when the letter "m" was left out of methamphetamine, this was a material change and the law was no
longer absolutely clear. The reasoning might be logical, but
the result (freedom for someone likely to harm society) is
not. The spirit or intent of the law, and the very assumption
on which law itself is based, got short shrift. (Oh, yes, it
could be argued that strict interpretation of the law protects
our freedoms; and in the same vein it could be argued that
no drug should be made illegal because eventually the government will end up curtailing our right to buy aspirin. I
don't buy the argument that taking one step forward leads
inevitably to falling off a cliff one mile away.)*

R.1.3 The Fallacy of Rigid Interpretations or Behaviors

In mathematics, one can multiply any number by another
and get a result. But mathematicians discovered that multiplying by zero could yield ridiculous results. For instance, 3
$\times$ 0 = 5 $\times$ 0; hence 3 = 5. So, they decided that one could
not perform certain operations with the number zero.
Results dictate the rules. Same thing in logic. Take, for
instance, the paradox of the self-reflective statement in the
rectangle below:

> All statements in this rectangle are false.
>
> A is C

---

*I suppose that judge Ellen Morphonius would not have freed the drug
manufacturer. An extremely popular Florida judge, she appeared on the
TV programs "60 Minutes" and "Cops." Her philosophy is that she tries to
follow the letter of the law as much as she can, to a point. As she puts it,
when the letter of the law "clearly violates common sense," she does what
common sense dictates, and too bad if the appellate court reverses her. To
me, she abides by the rule of reason, and behaves ethically.

If all the statements are truly false, the statement "all statements in this rectangle are false" is itself false, and "A is C" may be true. This is not a good result, particularly so if we know that A is not C. Hence, just as with multiplication by zero, logicians decided that the truth of a statement does not apply to itself. The ethical implication is that, for instance, one can be intolerant toward intolerant people without being intolerant. Results tell us whether the rules of math, logic, or ethics do or do not work. By the fruits we know the tree.

R.1.4 Is It True That the Ends Never Justify the Means?

Imagine that a terrorist is holding ten people hostage. In five minutes he will detonate an atomic bomb and wipe out a city with one million people. You can press a button and blow him up together with his innocent hostages and save the city. Would you press the button?

My hand might freeze, but I would make every effort to press the button. To me, under these circumstances, the killing of ten innocent people is justified by the saving of one million innocent people.

R.1.5 Not All Ethical Problems Have a Solution

Imagine now that pressing the button will blow up the above-mentioned terrorist and his ten hostages. But if you don't press the button he will blow up at most one house with only 10 people. If you know what to do, please increase or decrease the number of hostages or of people in the house, until you don't know. Thereafter add to the terrorist's location or to the house all the art works of Leonardo da Vinci, and see if you can balance this with a given number of lives. The ethical problem in the section above was relatively easy, because many lives were traded for few lives. The "currency" (lives) was the same. If the number of lives to be lost is similar whichever action is taken, things get complicated.

When the survival of precious works of art have to be balanced against one or more innocent lives, the "currency" is not the same and the problem may not be soluble.*

R.1.6 Some Ethical Solutions May Be Unethical in the Long Run; Mores May Be a Guide

Let us go back to the period from about 1000 BC to 1400 AD. Let us picture a very sick person, with visible skin sores, entering a village and begging for water or food. The villagers are relatively enlightened people, and react in an appropriate way. What was "appropriate" during that period of time? We might assume that immediate help was.

Well, the traditional and moral response was indignation! The poor wretch would have been driven out and away under a hail of stones, or worse. People with certain "abnormalities" (say, like leprosy) often posed a long-term hazard to the whole community. The comfort of one needy person was nothing compared to the risk to the health of the whole village. In those times, some forms of cruelty had an ethical justification.

The ethical effects of an action have to be weighed over the long run. Yet, it is very difficult to predict the future.

I tend to question every ethical or moral rule, among other things. But, unless I see very compelling reasons to do otherwise, I tend to follow convention. Old mores have been tested through time; more often than not they have good reasons to be what they are.

---

*If a democratic society freely agrees to place a monetary value on the life of children, breadwinners, older persons, etc., ethical solutions may be possible to the problem above. (See "Science, Values, and Human Judgment," by K. R. Hammond and L. Adelman [*Science* 194:389–396, 1976]. This is a landmark article on judgments involving scientific valuation.) Nowadays this and similar types of valuation permit compensation of survivors of air crashes and people who lose body parts; if and when practiced, they also permit a rational allocation of health and safety resources.

R.1.7 Flowers for the Taking

If we go to a public park, we may feel like taking some of its flowers to give them to someone we know. Even if the park had no sign prohibiting the taking of flowers, it would be unethical to take some: If everyone would do as we do, there would be no flowers left for anyone to enjoy.

But, what if we were out in the wilderness, in land that belonged to "nobody"? Fifty years ago, I would have said cutting a few flowers would present no ethical problems. The few people who would visit the same area couldn't take more than a fraction of the flowers. Today, with off-road vehicles and lots of people everywhere, one would have to be careful: The flowers might be those of endangered plant species, or the flowers themselves might be food for endangered insect or animal species. So, unless we know that the flowers are from a common type of plant, we should not take them. Otherwise we might be causing extinction, an irreversible outcome.

R.2 ETHICAL DECISIONS IN SEPTIC SYSTEMS PRACTICE

Now comes the fun part. In the following subsections I use true stories to describe ethical problems in septic systems practice. From the clues given in the previous pages, you might be able to guess what the solutions were (i.e., the "outcomes"). You can also figure out your own different solutions to the same or similar problems, and define and analyze the assumptions that give rise to the different solutions.

To preserve anonymity, CSSS stands for a county septic systems specialist; and SSC stands for a (private) septic systems consultant.

R.2.1 Ladies in Distress

In the late 1970s, a county septic systems specialist, CSSS, visited the house of two elderly ladies who had a failing leachline. Their lot was very small, and there was no room to install the required and necessary length of replacement leachline. The ladies were barely managing with Social Security benefits, so they could not afford the installation of a holding tank and its expensive frequent pumpings. A consultant had told them that there was nothing he could do for them. They were faced with vacating their home by force of law.

The policy CSSS had inherited from a departmental supervisor was that he should not do any engineering type of work for a private party: His county could be sued if things did not work well. He should just stick to enforcing rules and regulations.

Analysis

CSSS felt that the county had a moral responsibility to do something if something could be done. After all, the county had allowed people to build houses served by a system that would fail. CSSS knew that he could help the ladies, but only by "bending" the codes and doing some "engineering."

Depth to groundwater was not a problem. Though a per-code length of leachline would not fit in the small lot, the same amount of absorption area could be obtained with a shorter but deeper leachline: Instead of the per-code maximum 3 feet of gravel below pipe, CSSS could give credit for the sidewall absorption area surrounding 5 or 6 feet of gravel depth below the leachline perforated pipe. This deeper type of leachline would work very well, so the probability of a lawsuit against the county was reasonably low.

Outcome

CSSS prescribed and authorized the deeper leachline, and instructed the ladies on how to take care of their septic system and reduce the amount of their wastewater. To this day, there has been no lawsuit.

In 1986 a new County Administrative Officer assumed command. He sent a letter to every county employee. In this letter he stated that an employee's mission was not to enforce rules and regulations, but to help people. CSSS's action was vindicated.

R.2.2 Allowing Illegal Installations

CSSS heard that C, a contractor, was installing illegal septic systems without a permit: Instead of installing seepage pits per code, he would dig a hole with a backhoe, place a vertical perforated pipe in the hole, and fill the space between the pipe and the soil with gravel. It made sense. Through an intermediary, CSSS contacted C and let him know that CSSS would keep confidential everything C would reveal, including C's identity. C trusted CSSS, and they had a talk. C was operating illegally because his installations violated the code; however, they were working very well. His clients were retired people who could not afford the per-code seepage pits (about 30–40% more expensive). CSSS offered to try to legalize his type of installation, and C agreed.

CSSS discussed the problem of legalization with the county Chief Plumbing Inspector. This individual felt that he could not authorize something that did not meet the Uniform Plumbing Code requirements. (Ah, the danger of lawsuits!) And he asked CSSS to reveal the identity of C, the contractor.

Analysis

Per conventional morality, one's word is sacred. The contractor trusted CSSS's word, and CSSS could not betray his trust. The installations were working well, and nobody was being harmed. On the other hand, CSSS was a county employee, and owed allegiance to the county authorities.

Outcome

CSSS told the Chief Plumbing Inspector that he had given his word and that he could not betray the trust of the contractor. The Chief understood. Things went back to "normal."*

R.2.3 The Power of Tears

CSSS reviewed a perk report for a lot with problematic soil conditions on which a house was partially built. CSSS told the owner of the property that he saw potential problems with a septic system installed on that lot, and that he could not approve it. The lot owner's wife started crying bitterly, and said that she had umpteen children to support (and CSSS did see part of her numerous brood), and that the sale of the house was their only income. On one hand, CSSS was moved and wanted to help. On the other hand, CSSS felt that it was not ethical to approve something he wasn't sure would perform, though liability rested with the engineer who prepared the perk report.

---

*If I were a psychiatrist or a lawyer and a client told me he would go out and murder someone he did not like, I would feel free to betray his trust. In my view, ethical rules and protections need not be extended to those who break them by a most unethical and irreversible act such as murder. If the expected unethical act were reversible and far less drastic, like damaging property, I might have kept mum: It is to society's benefit that professional confidentiality be respected, in most cases.

Analysis

Being ethical in the sense of helping people in distress could end up being unethical to the future buyers of the house. If the owners had been the average speculator or developer, denial would have been no problem. But, given the circumstances . . .

Outcome

CSSS escaped the quandary by telling the woman that, though he could not approve the perk report, she could appeal to the top decisionmaker, her county supervisor. (CSSS broke with policy, because he should have referred her to someone directly above him. But CSSS suspected that the person above him or even the person above this person would just enforce the rules and regulations, and CSSS wanted to give the woman a chance.)

Weeks later, CSSS's supervisor told him that a woman "cried her heart out" at the office of a county supervisor, and that her perk report had to be approved. He knew how CSSS felt about principles, and to spare CSSS any embarrassment he kindly offered to approve the report under his own signature. CSSS had no qualms about signing it himself: a decisionmaker had decided for him.

R.2.4 Duty of Civil Servants

There are instances when what a CSSS is expected to do or say violates CSSS's perception of what's right to do or say.

Analysis

As I see it, civil servants are paid by and serve the taxpayers. Though final responsibility rests with the ultimate decisionmakers (up the chain of supervisors) who should usually be obeyed, civil servants are protected by civil service rules so that they won't have to obey orders they know could damage a community or violate ethics. This is a great privilege. However, there is a fine line between being a rigid SOB and a principled civil servant.*

Outcome

Personally, I was very fortunate to have worked for a rather "clean" governmental entity. I am acquainted with a very able, honest, and principled colleague whose life was made so miserable in another jurisdiction that he had to quit his job. My advice to prospective public employees is to do as they are told for as long as they are on probation, and to be ready to change jobs if they can't improve things that are patently wrong.

R.2.5 The Solar Power Plant

A large solar power plant development project had cleared the environmental impact report procedure and hearings, but its perk report was holding up its final approval. The soils at the proposed development site were terrible. They would not have passed under county codes and regulations. CSSS was on the spot.

Analysis

There was no doubt in CSSS's mind that the benefits of such a unique demonstration plant to society were extremely high. The site's remote location and isolation minimized con-

---

*The advice of some experienced and highly ethical acquaintances is as follows: If a request comes from a city council member (or someone else who has been elected and hence should be obeyed), one should make sure that the majority (3 out of 5) of the council members concur. If the request comes orally from a city manager or administrator, one should reply with a letter stating "I understand you have requested such-and-such." Then, one should keep a copy of this letter in one's own home: Locked office desks and files can be ransacked in the dark.

cerns about public health hazards, and its extremely large area probably had spots suitable for installing the required small septic system. Depth to groundwater was not a problem. In the past CSSS had authorized use of open-air sewage lagoons in remote areas, so he had a precedent to go by.

Outcome

CSSS actively advised the perk consultants what to do and what to look for. CSSS forgot about codes and regulations, and concentrated on designing a system that would work. He told the project's perk consultants that he would accept deep and/or long leachline trenches; the trick was to excavate as much soil as necessary to intercept a maximum amount of permeable strata. One leachline would be long and not too deep, another, maybe shorter but deeper than the code's maximum 3 feet of gravel below perforated pipe. In the unlikely event that in 10 years or so a leachline failed, the consultant would have to evaluate and report this event so that the replacement leachline could be sized to last indefinitely.

R.2.6 Bribes

Of course, the dilemma is not whether to take bribes or not, but how to refuse them. If bribe offers are refused with indignation, the briber might learn a lesson; if they are refused graciously, the briber might not.

CSSS's first encounter with an attempted bribe took him by surprise. He was inspecting a leachline installation, and the installation contractor gave him a large-denomination bill for "coffee money." CSSS was, as usual, dressed like a slob (i.e., comfortably); so he thought that maybe the contractor took pity on him. Perhaps. Anyhow, CSSS did not want to embarrass the contractor. He replied "Thank you, I don't need it. I am well-to-do."* And CSSS insisted he did not

---

*His reply wasn't quite truthful, but it was ethical.

need the money. The contractor took the bill back with a funny smile. Months later, another contractor told CSSS about the incident, so the original contractor must have passed the word around.

Many years later, CSSS was offered money to "take short-cuts" or to render favorable opinions. His answers were something like "No, I can't," or "Don't waste your money, I'll do my best regardless."

CSSS felt he didn't have the heart to lecture such people about ethics. Most likely, doing so would not affect their future behavior. Probably they learned from experience that bribing is the way to get ahead in a competitive world. CSSS preferred to be polite when refusing offers.

R.2.7 Private Consultant Versus Society

I have met consultants who will push for their client's interests, without regard as to whether their client is right or wrong. I feel it is a matter of principle to do the utmost for one's client, but only up to a point. If I hired a consultant, I would expect the same from him/her. I feel that there is a point where it would be unethical to do the client's bidding: Professionals of various kinds are part of, and are licensed by, the society they live in. Ethically, this entails a responsibility to society.

The following sections illustrate the conflict between what's owed to society and what's owed to clients.

R.2.7.1 Advocacy

A consultant requested SSC's help. A certain jurisdiction would not approve his client's housing development on the grounds that its density was too high. Because of certain peculiarities of the project, SSC could read the codes in a way that would allow the development to proceed. It was a tricky interpretation: It followed the letter of the codes, but not their spirit. It was biased.

Analysis

SSC felt it was a matter of principle to help those who came to him for help, but also felt uneasy about helping his client. He reasoned that:

1. It is ethical for trial attorneys to present only one side of a problem, the one that favors their clients. He could do likewise.*
2. He had told the staff of the jurisdiction involved that, as a public service, they were free to consult with him at no charge. The jurisdiction staff had done that once before. If the staff would call him and ask for his opinion about the development at hand, he could speak his conscience and alert them to the biased interpretation.
3. No significant irreversible damage to groundwater would result from the approval of this project alone, or a few more like it. This was a crucial point when SSC decided to help the consultant.
4. The public (or, more exactly, some members of the public) needed housing projects like this.
5. SSC would tell the consultant that the solution involved bias; it would be the consultant's decision whether to promote his client's interest or not.
6. Since he wasn't absolutely sure whether he was doing the right thing, how could SSC make sure he wasn't selling his integrity out? There was one way: definitely not accepting money for the advice he would give.

Outcome

SSC revealed to the consultant how to interpret the codes in a way biased in favor of the developer. The consultant did.

---

*I think that SSC erred by trying to imitate trial attorneys. First, I feel uncomfortable with "tunnel vision" positions, within or outside the halls of justice. And second, the comparison with trial attorneys might be fallacious: though some people murder other people, this doesn't entitle one to say, "Why can't I do that, too?"

The jurisdiction staff did not consult with SSC. The development was approved.

SSC refused any money from the consultant and from his grateful client.

R.2.7.2 Using an Inaccurate Methodology

The chief engineering geologist of a consulting firm hired SSC to prepare a perk report for seepage pits in a location with problematic soils. SSC told him that there was no accurate way of predicting pit longevity with the type of stratified soils at the site. The geologist confided that he too was often faced with situations that involved decisions well beyond state-of-the-art knowledge, and that one had to do the best one could, and hope for the best.

Analysis

First SSC wondered if he should take the job at all. Though he did the ethical thing and did not present himself as a know-it-all, he was still bothered by the use of pits in complex stratified soils. He was aware that, in a large portion of the county that had simple stratified soils, pits worked well for many years. He was also aware that in another portion of the same county that had complex stratified soils, there were problems: The county started receiving reports of "groundwater" at about 30 feet, though groundwater was over 100 feet deep: The cumulative discharge of many pits in the area was too much to percolate through one or more of the limiting clay layers below 30 feet. (Lateral saturated sewage flow is a no-no, as pathogens do not get filtered-out efficiently.) SSC also knew that much of the county's desert region was theoretically unsuitable for septic systems; yet much development had already occurred in it, and only a few areas were having problems. One could not go to the Board of Supervisors and argue that because of the theoretical possibility of problems, practically all growth on septic systems had to be stopped. County supervisors, the decisionmakers, were pro-growth and undoubtedly felt that development was worth the risk.

For at least the previous 13 years, other consultants had been doing perk reports for pits in the problematic area SSC was to tackle, and they were using the inaccurate but required "falling head" methodology. If SSC did not take the job, someone else would; and by merely following rules, that someone else would do much worse than SSC could. If SSC did take the job, he could learn and pass on what he learned to the jurisdiction.

Outcome

SSC took the job. With the help of the chief engineering geologist and his staff, SSC prepared the most extensive and thorough perk report ever submitted to the jurisdiction for the area involved. Since he was forced to use the "falling head" test, he performed it. But he noted in the perk report that this type of test was flawed. Basing his suggestions on hunches rather than on the test results, SSC recommended roughly 3 to 5 times more pit absorption area than the minimum county requirements. And SSC prepared instructions for the prospective users of the pits, to explain to them how the septic system operated, and how to take care of it, and its limitations.

(Months later a pit installer very experienced in the same area with the problematic soils told SSC that he usually recommends 3 to 4 times more pit absorption area than the county-required minimums. Whew! SSC had guessed right!)*

---

*I feel that one cannot change the world by oneself and at once, and that one need not abstain from doing as other ignorants do, if one can do at least as good a job, poor as it may be. But, to be ethical, I also believe that it is important to take a stand and point out things that aren't right to those who can correct them.

R.2.7.3 Cumulative Impact

The groundwater in an area was degraded with nitrates, and continuing development on septic systems was making things worse. Local jurisdictions tried to slow down development on septic systems by requiring the filing of a "nitrate impact report" to prove that each development by itself would not affect groundwater. Developers came to SSC for help.

Analysis

If SSC did his best to help his clients, he could show that each of their individual developments, by itself, would not impact the groundwater. But he felt that he would be helping to exacerbate an areawide problem.

Outcome

SSC's solution consisted of:

- writing reports on the basis of explicit assumptions (which turned out to be favorable for each separate development)
- pointing out to the jurisdictions involved that their objectives had to be properly defined
- offering free consultations to help out the jurisdictions in their attempts to protect groundwater

It was like picking a flower in an open field, and then trying to induce the owner of the field to post a sign prohibiting the picking of flowers. SSC's individual clients would not be harmed. Future developments might be stopped, and SSC would deprive himself (and other consultants) of similar consulting jobs, but groundwater and the communities that depended on it would be better protected.

R.2.7.4 *Civil Servant Versus Private Consultant*

SSC wrote a perk report and submitted it to CSSS. The report claimed that lots were suitable for septic systems; but the raw data presented contradicted this claim. SSC told CSSS that some of his clients would not pay him if he were 100% truthful and wrote in his perk reports that septic systems would not work on their lots.

Analysis

SSC was being honest with CSSS, and with the public CSSS represented. If CSSS insisted on SSC being 100% truthful, CSSS would either force SSC to be dishonest or deprive SSC of the type of clientele that would hire a less honest consultant; then less honest consultants would gain and honest ones would lose.

Outcome

SSC and CSSS reached an agreement: In his reports, SSC would give CSSS all the accurate field data needed for CSSS to form his own opinion, and SSC would be free to praise the suitability of any site he wanted even if the data indicated the site was unsuitable.

SSC did his part, and so did CSSS. When the time came, CSSS was the one who gave the bad news to SSC clients.

CSSS didn't feel empathy for clients who would not pay an honest fellow. CSSS wished consultants would not have to use a subterfuge to get paid, and that all consultants at all times would give society's interests top priority in their reports, as implied by, say, the American Society of Civil

Engineers code of ethics.* But with the world as it was, CSSS thought that it would be unrealistic to expect such idealistic behavior to materialize anytime soon.

---

*A county Environmental Analysis Unit solved this problem thusly: It itself hired environmental-impact consultants with the developers' money. If hiring by a public agency can be kept honest and fair, this solution seems pretty good. Recently I had a conversation with a very ethical and able public employee who reviews geotechnical reports for a Water Quality Control Board regional office. He was upset by the fact that developer's consultants were submitting to him reports that (in his words) "came within an inch of lying"; their interpretations and conclusions were always extremely favorable to their clients, even if their own field data made a strong case for just the opposite type of interpretations and conclusions. He felt that these consultants were not behaving honestly. I felt that if the field data were honest and permitted him to reach opposite conclusions with ease, the consultants did a decent job: They had to behave as all other consultants were behaving or go bankrupt. When consultants can be disciplined for acting deviously, then I'll feel differently.

Appendix S

SCIENCE, LAW, ETHICS, ECONOMICS AND A NITRATE POLLUTION CONTROL SAGA

On March 30, 1989, a special CBS broadcast, "L.A. Justice," informed us that the Los Angeles county courts—the largest in the nation—were congested by too many lawsuits to the point of "pandemonium" and were highly inefficient. The same day, KCET-TV ("California Stories") broadcasted an interview with a lawyer who represented the American Civil Liberties Union (ACLU) in its fight to solve the courts' problem. The ACLU's solution to this congestion problem was to sue the courts!

Perhaps lawyers are prone to thinking that every problem has a solution: a lawsuit. (What else?) Perhaps legislators are prone to thinking that the solution to every problem is a new law. (What else?) Perhaps psychologists might be biased in favor of psychoanalysis for everybody. (Of course!) As for myself, I admit having a strong bias, if this is a bias: I believe that, if a problem has a solution, the optimal path to an optimal solution is the scientific method. (What else?)

Essentially, the scientific method consists of defining and solving a problem by means of *answerable* questions. One must define a problem in such a way that the variables of interest can be observed and measured *under a proven or acceptable methodology that can be monitored and improved*. Take the famous Byzantine discussions about how many angels can dance on the head of a pin: The discussions lasted forever and led nowhere because no one was able to measure directly (under a commonly accepted methodology) or indirectly (with a mathematical model of an angel's relative size and anatomy) how much room an angel needs in order to perform a specific dance in a defined way.

In Appendix M, Section 1, Perkins[1] was quoted regarding the dismal results of legislation that induces even more legislation and so on. I see this spiral of legislation as the consequence of laws that ignore everything about science but its trappings. (And we should not expect much improvement in this regard. This problem is likely to get much worse before it gets better: The proportion of college freshmen wishing to major in math and sciences fell by half—to 5.8%—between 1966 and 1988.[2]* Hence, our future legislators are likely to be elected by—and from—a population that is becoming more "scientifically illiterate.")

Local attempts to control the problem of nitrate pollution in groundwater serve to illustrate interactions of flawed legislation (and legal requirements), technical knowledge, economics, and ethics. Because the subject is highly specialized, I have placed it here, as an Appendix. But, I have no doubt that the events described are commonly replicated in (and applicable to) many other fields.

The major actors of the story that unfolds below are "good guys" who try to prevent pollution of groundwater the best way they can with their limited resources:

---

*The reason for this trend may be that the income of engineers (and scientists) is less than half that of lawyers, per the September 25, 1989 issue of *U.S. News and World Report* ("Best Jobs for the Future"). Commenting on this statistic, Mr. M. Jacobs (*U.S. News and World Report*, October 23, 1989, "Letters to the Editor;") noted, "The best minds are going to corporate or tax law . . . producing nothing more than advice . . . on how to circumvent the travesty of laws that they themselves created. Until we get a grip on a legal system out of control, we'll continue to slide in competitiveness, while the brightest among us make fortunes giving advice on how to sue one's neighbor."

- sanitarians (environmental health specialists) in county jurisdictions
- engineers (and geologists) from a regional office of the California Water Quality Control Board (WQCB)

Other "good guys" are entrepreneurs called "land developers" or just "developers." They make a living by satisfying the population's need for housing at the lowest cost dictated by market demands. The problem is that *the lowest-cost development often depends on the use of septic systems, which generate nitrates that pollute groundwater.* Groundwater is the main source of potable water for much of southern California's population. The California WQCB must protect groundwater from pollution (per the California "Porter-Cologne Act"). But its hands are tied by legislation. In particular, it cannot abate pollution caused by agricultural fertilizers, which are a major source of nitrate pollution. And California Water Code section 13280 expressly prohibits the WQCB to abate pollution unless it can prove that there is a problem. Once people are dying left and right, it is usually easy to prove that there is a problem. Before then, it is very difficult, considering how the judicial system works. And by the time the proof is at hand, it is often too late to repair the damage.

Now, a chronicle of happenings.

Around 1987–1988, a regional office of the WQCB accepted two methodologies to evaluate whether residential development impacted groundwater with nitrates. One methodology was briefly mentioned in *Septic Systems Handbook* and is elaborated in Appendix U, Problem No. 12. The other one was presented by N. Hantszche at the 31st annual meeting of the California Directors of Environmental Health in 1986. It aims to find out how much area is required (per each dwelling served by a septic system) in order to catch and percolate enough rainfall to dilute an assumed sewage percolate concentration of 40 mg/L of N (nitrogen) to the drinking water standard, 10 mg/L (equivalent to 45 mg/L of nitrate). Wherever there is sufficient and regular rainfall and percolation, as in some states east of the Mississippi, this approach should be useful.

Literature prepared by the WQCB staff described and applied Hantszche's procedure as follows. My comments are in brackets.

Most studies dealing with areas less than ideally suited to septic systems recommend densities closer to one [single house] system per five acres. Average septic system density in the XXX area is one system per 0.17 acre.

. . . While a number of factors . . . affect the nature and degree of the contamination problem caused by septic systems, the density of the systems is the principal controlling factor. As lot size increases, groundwater contamination problems decline since more groundwater is available for dilution of the septic tank effluent percolating into the aquifer.

Using the following equation, it is possible to estimate the critical development density (Dc), defined as the acre/dwelling unit ratio, that will result in an areawide percolate nitrate nitrogen concentration of 10 mg/L (45 mg/L as nitrate).

$$Dc = (2.01)(Np - 10)/(DP)(10 - Nb)$$

Dc = Critical development density, acre/edu. [An edu is an "equivalent dwelling unit," a single house or something like it.]

Np = Wastewater nitrate nitrogen concentration, mg/L (assumed to be 40 mg/L in the following calculations).

Nb = Background nitrate nitrogen concentration of percolation rainfall (mg/L); assumed to be 0.5 mg/L.

DP = Deep percolation of rainfall (in/yr); 12 inches/year assumed.

2.01 = Conversion factor for assumption of the discharge rate of 150 gallons of effluent/edu/day.

[At this point I would like to note that 150 gallons of effluent per day with 40 mg N/L results in a nitrogen output of 8.3 kg/year/edu. This is low. The 40 mg N/L figure is probably correct as a local average, but the average outflow in neighboring sewered areas is 280–300 gallons/day. At 300 gallons/day, the nitrogen output would be 16.6 kg/year/edu, similar to the figure in Chapter 12 of this book (Section 12.1). This book uses 6 kg N/person/year; since there is an average of 3 people/edu locally, an edu would discharge 18 kg N/year; septage pumpings and some denitrification would reduce this figure by roughly 20% to about 15 kg N/year.]

Therefore,

$$Dc = 2.01 (40 - 10)/12(10 - 0.5) = 0.53 \text{ acres/edu}$$

The value of 12 in/year for deep percolation of rainfall assumed in this calculation is probably unrealistic . . . a more realistic figure (3.72 in/year) yields the following result for the study area:

$$Dc = 2.01(40 - 10)/3.72(10 - 0.5) = 1.71 \text{ acres/edu}$$

Based on these calculations, septic system densities of 0.53 to 1.71 acres/edu should not result in violation of the primary drinking water standard for nitrate in ground water. Currently, the average density of developments on septic systems in the study area is 0.17 acres/edu.

[The assumption that 3.72 inches/year percolate down to the aquifer in the local area under study is probably incorrect. Per Bulletin 33 of the California Department of Public Works, 1930, all of the scant local precipitation, about 16 inches/year, runs off or is lost to evapotranspiration. I pointed this out to a WQCB engineer and he noted that the suspect figure was derived in an engineering study; I checked this study's tables and saw that the suspect figure was an average for a very large region, which comprises a small area with much precipitation—mountains—and a very large area with very little precipitation—desert! Though the WQCB calculation might be valid for the region as a whole, it has no scientific validity for the local area "XXX." "Estimation of Natural

Groundwater Recharge"[3] emphasizes that, regarding groundwater recharge in arid areas, "No single comprehensive estimation technique can yet be identified from the spectrum of methods available; all are reported to give suspect results." And the WQCB engineer was aware of the fallacy of averages,* but as an engineer, presumably he had to act and do the best he could with the data available.]

Well, since most local developers were placing single homes on lots smaller than 0.53 acres, the developers' engineers had to find another way of justifying such development to the WQCB. And they found it in Chapter 12 of *Septic Systems Handbook* (Section 12.3). So did a consultant we'll call SSC. But after he produced many nitrate impact reports, SSC felt bad. He was aware that piecemeal approaches don't work too well, and the impact studies were piecemeal. It just didn't feel right to continue with the same routine.

Furthermore, a county jurisdiction was requiring the following from developers:

Prior to recordation [of the tract] submit an engineering report assessing the impact of the subsurface septic system in relation to nitrates on the underlying groundwater. The report should include geologic and hydrologic evidence that the discharged waste will not individually or collectively, directly or indirectly, adversely impact the water quality (nitrate level) of the area. Should the water quality be adversely affected, hook up to sewers will be required.

When SSC called this jurisdiction, a sanitarian told him that "just any report would do"; the sanitarian didn't understand what the requirement entailed. Its words were clear, but no one could have understood the requirement, not even whoever authored it, because it had not been expressed per

---

*An easy way to illustrate this type of fallacy follows. A medical researcher visits a tribal chieftain. One of the chieftain's four wives is pregnant, in her eighth month. Since 8 (months) divided by 4 (wives) equals 2 months per wife, the reseacher reports to the chieftain that his wives are two months pregnant.

scientific methodology. If it had been, the term "adverse" would have been defined in a quantitative way and in function of a methodology or mathematical model. This in turn would have defined all the variables that must be estimated or determined. Fortunately, the sanitarian knew one criterion for significance, "an increase of 1 ppm (part per million) nitrogen content in the water." This enabled SSC to write a few more impact reports.

The sanitarian's jurisdiction was under the influence of the WQCB. SSC wrote a letter to the WQCB engineer, and suggested that:

> . . .By determining what is the carrying capacity of the aquifer for added nitrates, it might not be necessary to require developers to produce a nitrate impact report every time they want to record a tract. . . .

After sending the letter, SSC met with the WQCB engineer and county sanitarians. They discussed the issues and the difficulties of defining and measuring nitrate impact. For instance, the engineer brought up the point that, according to his geologist, the aquifer flow velocities SSC was using in his nitrate impact reports were too fast; a county geologist who was present stated that they were too slow. SSC pointed out that the velocities were obtained by a hydrogeology professor who tested 400 wells in the impacted area, and that this was the best that state-of-the-art knowledge could offer. The WQCB engineer stated that he would not require nitrate impact reports in the future, because he thought that it was easy for consultants to obtain favorable results by making subjective (but not "incorrect") assumptions. He would probably require a minimum lot size of about 0.5 acres per home.

At this meeting, SSC might have misled the county sanitarian when he told him that the way to make results objective was to define in advance the parameters needed and also to make the parameters restrictive to ensure safety. The sanitarian only did the latter and restricted the allowable depth of mixing in the aquifer to 100 feet maximum. Under this restriction SSC could no longer prove that the average subdivision tract would have a nonsignificant impact, if he continued using the methodology described in *Septic Systems Handbook* Section 12.3. But SSC knew he could achieve the same result with a more complex methodology. (This methodology is described in Appendix U, Problem 14.)

When SSC prepared another nitrate impact report for a new client, he developed a more complex and realistic model. He submitted it to the county. The county approved it and rescinded the requirement for nitrate impact reports: It appears that the report made it easy to visualize why even a high-density subdivision, by itself, over an aquifer hundreds of feet below ground, would not cause a significant nitrate impact.

Later on, at an official meeting, a WQCB engineer met with a group of consultants and explained that the WQCB would impose a minimum lot size of 0.5 acres for single residences. SSC mentioned that the calculations to derive this lot size were based on the fallacy of averages (averaging a large area with little or no rainfall recharge and a little area with much recharge) and that specific communities might not be protected: The groundwater of some communities could be severely impacted by nitrates at the same time that other neighboring communities would be experiencing no problems at all. SSC also mentioned that since the amount of percolated precipitation reaching groundwater was probably zero throughout most of the area of concern, lot size was not as relevant as it appeared. Within a given width perpendicular to the flow of an aquifer, 50 homes on 0.17-acre lots would result in a groundwater nitrate level nearly identical to that of 50 homes on 10-acre lots. The engineer was quite aware of this. He said that the WQCB had already commissioned a (nationally known) engineering firm to do a detailed areawide hydrogeological study. This was a movement in the right direction.

However, a movement in the right direction may not be enough. I doubt that a solution will be found anytime soon, unless the budget for the hydrogeological study is very generous or else a totally different way of thinking and problem-solving is adopted. I have worked for and with engineers and reviewed thousands of their reports. It is my impression that consultant's reports tend to reach conclusions not so much on the basis of what's really happening out there in nature (which is very difficult to ascertain), but very much on the basis of legal and budgetary constraints. As they must, they do the best they can with the money and data available. Often, this is good enough; sometimes, it isn't. If their conclusions are based on flimsy data or assumptions, so that the conclusions are worthless, that's the best they can do; they can't work for free. And I don't fault them for this. But much too often such reports are treated like God's word by less sophisticated professionals or nonprofessional people, and they can lay the ground for monumental goofs when translated into policy, legislation, or planning schemes.

And there is another troublesome consideration. In general, consultants give their clients what their clients request. If the clients don't know what they want, they don't get what they need. I suspect that the WQCB doesn't know what to require because the decisionmakers, the legislators who write the laws, don't know either. For example, what is the definition of "adverse impact"? Is it to be defined as reaching a concentration of 1% or of 100% of the standard (45 ppm nitrate)? Timewise, will the concentration be measured as an average or on the basis of data obtained during the worst three consecutive drought years? How is it going to be defined regarding space coordinates? Is it going to be based on the degradation of the upper 10 feet of aquifer or of the lower 10 percent of well depths? Is it to be based on a percentage of wells protected, of well yield protected, or of water consumers protected? What depth and lateral spread of pollution are to be considered adverse impact? Which methodology for sampling and analysis will be chosen? At what cost does a certain level of pollution control effort become excessive? Beyond what level of expenditure would money have more utility if spent on controlling other pollutants, on education, on providing a reverse-osmosis device at every home, or on something else?

Though the last chapter of the nitrate pollution control saga will not be written by me, and perhaps not by anyone in this century, I hope that this narrative yielded insights regarding some of the ways in which technical/scientific knowledge (or, more accurately, the lack thereof) may interact with law, policy, economics, and ethics, among other things. Everything is related to everything else, and problem-solving ought to take this into account. A similar saga could have been written about a myriad of other problems.

We have seen that an ethical concern, protecting the quality of (drinkable) groundwater, was embodied in legislation. Possibly another ethical concern (preventing abuse of power), resulted in legislation that tied up at least one hand of the WQCB. Then we saw how some jurisdictions attempted to control an improperly defined nitrate pollution problem the best they could with costly trial-and-error efforts. Ironically, while the present requirements for larger lot sizes in our arid areas have a very weak scientific basis, such requirements might do some good because of the "back-door" impact of economics: the cost of larger lot sizes may induce developers to build on small, sewered lots. And sewering and treating (and denitrifying) sewage at a sewage treatment plant can reduce the nitrate pollution problem.

REFERENCES

1. Perkins, K. 1989. California environmental policy: a time for change. *Calif. J. Environ. Health* 12:16–18.
2. Green, K. C. 1989. A profile of undergraduates in the sciences. *Am. Scientist* 77:475–480.
3. Anderson, M. 1989. Review of *Estimation of Natural Groundwater Recharge*. *Am. Scientist* 77:491–492.

Appendix T

DEALING EFFECTIVELY WITH BUREAUCRACY

One would imagine that professionals are hired to make, understand, and adapt or bend rules for sound ethical or technical reasons. But when professionals become part of almost any organization, public or private, they may not act as one might expect. An organization's bureaucrats are human beings who try to do their jobs, safely. If they are subservient to "the rule of rules" and follow the letter of the law (or ordinance or rule), they are very safe (even though the results of their actions may sometimes be irrational or counterproductive). If they follow the rule of reason and bend or break rules, they may eventually goof and be subjected to reproof and chastisement. (That's how the system works.) So, the easiest way to deal with an apparently unreasonable bureaucrat is to help this person help you.

Years ago, CSSS, a county septic systems specialist, was bothered by his county's requirements for expensive percolation test reports in some communities that had good sandy soils and didn't need such reports. To help prospective home builders, CSSS prepared an official form and titled it, "Soils Evaluation." When the soils were good, CSSS confirmed this fact by inspecting exploratory trenches in the field and filled out the "Soils Evaluation" form with a description of the soil profile and with a percolation rate that he assigned. This filled-out form substituted for the formal and expensive percolation test report. Home builders saved quite a bit of money because of this form. But, all of a sudden, one day an engineer from a state agency rejected these "Soil Evaluation" forms. He told CSSS that his state agency could not accept the forms because his agency's rules required filing a percolation test report.

So, CSSS immediately changed the title of the form to "Soils Percolation Test Report," re-filled it with identical data, and re-submitted it. The engineer (and his agency) accepted this "percolation test report" and the ones that followed. (This engineer was a very intelligent professional. He was aware of what CSSS did. Rules had to be obeyed, and he did obey. If he hadn't, his agency might have been sued by some punctilious lawyer, and he would have been in trouble. Bureaucracies tend to be highly risk-aversive.)

If one is at odds with a bureaucrat regarding the interpretation of regulations, one should politely and amiably request to have a talk with a supervisor. If the matter of disagreement is so fundamental that the regulations themselves should be changed, it pays to go directly to the person at the very top of the organization. Such a person is less bound by rules and regulations, and much more likely to follow the rule of reason. Before going to the top person, you should explain your point of view and reasoning to the bureaucrat, amiably. If you engage in what is perceived as unfair intimidation or power plays, bureaucrats may fight back in a way reminiscent of guerrilla warfare.

APPENDIX U

Problems Solved, Questions Answered

The following problems and questions give tips on how to handle obstacles one may encounter in professional practice. Some are based on errors committed routinely by expert consultants.

PROBLEM 1. Perk times of a mixture

We have a large mound of a dry uniform sand. It weighs one million tons. A perk test on this sand gives a perk time of 1 minute/cm. We mix and remix this sand top to bottom and side to side, and retest. We still get 1 minute/cm.

Nearby we have a large mound of a dry uniform gravel. It also weighs exactly one million tons. Its perk time is 0.05 minute/cm, both before and after we mix it as in the paragraph above. The mineralogical composition of the sand and the gravel are the same. The sand particle density is identical to the gravel particle density.

We mix and remix the sand and the gravel together, until the mixture is perfectly uniform everywhere, and we test and measure its perk time.

Multiple choice question

What's the most likely value of the perk test result on this mixture, in minutes per centimeter?

a) The arithmetic mean of 1 and 0.05
b) The harmonic mean of 1 and 0.05
c) The geometric mean of 1 and 0.05
d) A bit larger than 1
e) A bit larger than 0.05
f) A bit smaller than 0.05
g) Other: explain

Answer

(Surprisingly, mathematically oriented people are likely to give the wrong answer.) One must visualize what's going on. Imagine water percolating through sand. The path of a cubic micron of water is fairly straight downward. Now imagine that there are big stones or gravel interspersed within the sand. The path of the cubic micron of water is no longer as straight as before. This cubic micron moves down until it hits gravel or a large stone, then it moves somewhat sideways and then down until it hits another, and so on: The path is more tortuous and longer than in the pure sand. The path X has increased, and the permeability decreases. So, the velocity of flow decreases and *d* is the right answer. Mathematical averages are nonsensical in this case.

When dealing with other soil-related problems, one should try to form a mental picture of what's going on in the soil.

PROBLEM 2. Applicability of Darcy's formula

An investigator took a tin can, drilled a little hole at the bottom, filled the can with water to a height h above the hole, let the water flow out of the little hole, and checked whether the rate of decrease in height h (rate of fall in the water level) was directly proportional to the heights h of the water level at different times during the fall. He conducted two initial trials, with two measurements each, and in each trial the time for h to fall from 16 to 15 cm was half the time it took to fall from 8 to 7.5 cm. He concluded that the rates of fall are directly proportional to the heights h, as predicted by Darcy's formula.

Question A

Did the investigator reach a valid conclusion?

Answer

No. Bernoulli's fundamental law of hydraulics relates the pressure head, the velocity, and the elevation of points along a line of water flow. It is expressed as

$$p + 0.5DV^2 + Dgh = constant$$

where p is the pressure head, D is the density of water, V is the velocity of water, g is the gravitational constant, and h is the elevation above an arbitrary level.

Let us visualize the same variables at two points, one within the can and one just outside the can: one point is within the can, at the top of the falling water column, and the other one is just outside the can, at the beginning of the little spout of water (which is coming out of the little hole in the can's bottom). Let the first point be 1, and let the other point (at the spout) be 2.

Per Bernoulli's law,

$$p_1 + 0.5DV_1 + Dgh_1 = p_2 + 0.5DV_2 + Dgh_2$$

Now p_1 and p_2 can be eliminated, as both are equal to the atmospheric pressure. The D's cancel out. We can arbitrarily locate the perforation at a coordinate point equal to zero so that h_2 = zero and h_1 is the height above the orifice. And simplifying and rearranging the leftover terms of the equation above, we get

$$V_1^2 - V_2^2 = 2gh_1$$

We know that the fall of the water level is extremely slow compared to the speed of the water coming out of the tiny perforation (the cross-sectional area perpendicular to the direction of the flow in the can versus that in the perforation could be in a ratio of 1000 to 1, respectively, or higher).

Then, let $V_2 = uV_1$ where u is a very small fraction.

Therefore, $V_1^2 - u^2V_1^2 = 2gh_1$; and rearranging,
$V_1^2 = 2gh_1/(1-u^2)$; or, $V_1 = \sqrt{2gh_1/(1-u^2)}$

and since one minus the square of a very small fraction is still about one, we have that the velocity of fall is directly proportional to the square root of the height of the water level. So, the water level rate of fall at any height h should be (square root of 2 =) 1.4 times larger than at half the same height.

Darcy's formula applies only to flow through a porous medium, like soil or fritted glass. And the medium cannot have large continuous macropore channels.

Question B

The investigator repeated his measurements, but this time he didn't just determine the velocity of fall at a "high h" level and then at half that height. He conducted a whole series of measurements over a range of h's in two new trials. His measurements are given in Table U.1. (These are real experimental values). Study the data in the table and see if you can explain why the investigator's initial measurements led him to the wrong conclusion.

Table U.1   New Rate-of-Fall Measurements in a Tin Can Perforated at the Bottom

h (cm)	First Trial Time (min:sec)	Interval (sec)	Second Trial Time (min:sec)	Interval (sec)
16	0:3		0:0	
		6		7
15.5	0:9		0:7	
		8		7
15	0:17		0:14	
		–		7
14.5	–		0:21	
		14		8
14	0:31		0:29	
		9		8
13.5	0:40		0:37	
		–		6
13	–		0:43	
		17		11
12.5	0:57		0:54	
		8		7
12	1:05		1:01	
		–		9
11.5	–		1:10	
		18		10
11	1:23		1:20	
		12		7
10.5	1:32		1:27	
		7		8
10	1:39		1:35	
		–		9
9.5	–		1:44	
		17		9
9	1:56		1:53	
		12		10
8.5	2:06		2:03	
		10		11
8	2:16		2:14	
		11		10
7.5	2:27		2:24	
		11		9
7	2:38		2:35	
		11		11
6.5	2:49		2:46	
		11		12
6	3:00		2:58	
		11		11
5.5	3:11		3:09	
		12		13
5	3:23		3:22	
		16		14
4.5	3:39		3:36	
		15		14
4	3:54		3:50	

Answer

If we look at the first-trial data, we see that the water level took 6 seconds to fall from 16 cm to 15.5 cm. And it took 11 seconds to fall from 8 cm to 7.75 cm. Halving the height resulted in halving the rate: 6/11 = 0.54. However, the first-trial data show that the measured times are "jumpy" because of low precision and accuracy. (Compare with the data from the second trial.) The fall in water level when h is at 16 cm is very fast, and during the first trial the investigator didn't read the falling h level precisely enough. By comparing the first-trial intervals at 4.25 and 8.5 cm, when the water is falling more slowly, we get the expected result, that the ratio of decrease is proportional to the square root of the ratio of h. The same correct result is seen throughout the second-trial data.

The didactic value of this problem is that the experimental setup and the variables were very simple, yet an investigator was misled by two initial trials. It may be far more difficult to interpret data obtained from experimental setups involving actual soils and far more complex physical and biological variables.

PROBLEM 3. Depth of ponding and absorption rate

As reported by Laak,[1] the Long Term Absorption Rate (LTAR) of a soil surface with ponded septic tank effluent over it is roughly proportional to the cubic root of the constant height (or depth) of the effluent over it. (At a given depth of effluent ponding, a soil surface should keep on absorbing effluent forever at the same LTAR.)

Assume that this cubic root mathematical relationship is fairly exact over a wide range of ponding depths.

Question

In a given soil, the LTAR measured over a horizontal or vertical soil surface is 0.2 gallons/ft²/day with ponding 4" high on the average. Assume that we are going to install a seepage pit or a deep leachline trench in the same soil, and calculate how much bigger the average LTAR is in the side-wall of the pit (or deep trench) full of septic tank effluent if the depth of effluent will be:

   a)   20 times 4"
   b)   80 times 4"

Answer

Whether the original LTAR was measured under a 4" depth or a 4' depth, we can call the original depth "unity" and the other depth 20 times unity (or 80 times unity). The total LTAR of all the square feet in a surface area one foot wide and 20 (or 80) units long (deep) is given by the integral of the cubic root of h (h = depth of ponding), from unity to 20 (or 80). And the average LTAR of this vertical surface of square feet running from a height of unity to a height h = 20 (or 80) times unity is equal to the above-mentioned integral divided by the height h.

$$(\int_{h=1}^{h} h^{1/3}dh)/h = (3/4 \times h^{4/3} - 3/4 \times 1^{4/3})/h$$
$$= 3/4(h^{1/3} - 1/h)$$

When h = 20 or 80, 1/h is negligible. Hence,

$$= 3/4 \times h^{1/3}$$

When h = 20 times unity, LTAR is $0.75 \times 20^{1/3}$ = 2 times bigger.

When h = 80 times unity, LTAR is $0.75 \times 80^{1/3}$ = 3 times bigger.

And 2 and 3 times 0.2 gallons/ft²/day gives respectively LTARs of 0.4 and 0.6 gallons per *average* square foot of side-wall per day.

References

1.   Laak, R. *Wastewater Engineering Design for Unsewered Areas*. 1986. Technomic Publishing Co., Lancaster, PA.

PROBLEM 4. Sizing absorption areas

Question

Can leachlines be sized accurately on the basis of pub-lished data?

Answer

No. Let us compare the sizing of leachlines to tailoring a custom-made suit.

Let us imagine that we go to the "Precise Tailoring" shop to buy custom-made, fine suits. The precise master tailor main-tains that the fit of his suits is superb because he takes pre-cise measurements, to the nearest micron. (Readers of femi-nine gender, please imagine going to a seamstress rather than to a tailor.) So, we are at the shop. Our measurements are being taken. We see that the tailor's measuring tape has 99 marks labeled centimeters, and at the end it has a micro-meter calibrated to read 1 cm exactly from 0 to 10,000 mi-crons (zero to one centimeter). To our consternation, we see that the tape is made out of rubber or something like that, because it shrinks or expands randomly to 1/3 to 3 times its average length. The microns might be very precise and accu-rate, but the centimeters might be off by up to a factor of 9.

Wouldn't we be skeptical if the tailor measured our waistline as "exactly 10.0007 cm"?

Would you buy a suit from this tailor?

No? If you have been using perk rates or perk times in your practice as a consultant or as a specialist in the septic systems field, probably you have been doing the same type of "precise tailoring." And perhaps you have bought and sold the emperor's clothes, too. Explanations follow.

Let us assume that in a lot or tract we obtain a replicable perk time measurement of 10 ±1 mpi (minutes per inch). In practice, unless consultants have been trained and do their testing in exactly the same way and during the same season of the year and find the same soil (and the same soil air and water content and distribution and temperature), such precision and replicability would be very unusual. Nevertheless, let's assume that we do obtain this accurate measurement.

Now we look up a table of required absorption areas. Such tables can be found in the EPA Manual,[1] or in Perkins',[2] Winneberger's,[3] or Laak's[4] books. The last two authors recommend about twice as much absorption area as EPA in the usual 1–60 mpi range. Hence, our "tailor's tape" might be off by a factor of two.

Now, let us consider that leachline longevity will be affected by (at least) 2 highly variable parameters: variability of sewage flows and variability in composition of flow (fats, BOD, SS). It is also affected by temperature, aeration, microbial composition, etc. Our "tailor's tape" could well be off the mark by a factor of 10. The absorption area tables in the above-mentioned references 1–4 are like the micrometer at the end of the tailor's tape.

But this micrometer itself is not too accurate either! It is also affected by the variables mentioned in the previous paragraph. And the meaning of its measurements is not at all clear. Something terrible happens when published research is translated into practical recommendations.

For instance, an investigator may determine that certain sewage at a certain temperature and ponding depth over a certain type of soil in a soil column of given depth results in a given long-term acceptance rate (or absorption rate) for sewage. But when this finding is translated into a table of sewage application rates, all we see in the table is that, to a real soil with a given mpi (or permeability constant), we are to apply up to a given amount of sewage per square foot. All the other important variables, and their effects, are lost in the translation. Problems 3 and 5 in this Appendix include experiments to emphasize the difficulties of interpreting research results and translating them into practical recommendations.

(Fortunately, though published sewage application tables are not as precise and accurate as they seem, they all reveal that absorption areas are fairly invariant. The absorption requirement for a 60-mpi soil is only about 3 times larger than that for a 5-mpi soil. One can err quite a bit in determining a soil's mpi and still come up with fairly decent recommendations, if one uses substantial safety factors.)

References

1. U. S. Environmental Protection Agency. 1980. Design Manual: Onsite Wastewater Treatment and Disposal Systems. EPA-625/1-80-012.
2. Perkins, R. 1989. *Onsite Wastewater Disposal*. Lewis Publishers, Inc., Chelsea, MI.
3. Winneberger, J. T. 1984. *Septic Tank Systems*, Vol. 1. Butterworth Publishers, Stoneham, MA.
4. Laak, R. 1986. *Wastewater Engineering Design for Unsewered Areas*. Technomic Publishing Co., Lancaster, PA.

PROBLEM 5. Comparing rates of absorption, water versus sewage

What follows is an actual experiment to estimate the relative absorption rates of tap water versus septic tank effluent.

Experiment

In the first half of 1987 I had an experimental leachline built on my property. It was 3 feet wide and had 6 inches of 1"-diam. gravel under the perforated pipe. The first portion of the leachline had about 7.5 ft² of bottom area. It was going to receive all the flow from a septic tank. The overflow was to go to a distant leachline. With a vertical slotted pipe I could monitor the liquid level. The soil was a porous loam that had tested roughly 0.5 to 5 mpi.

On July 19, 1987, at 10 a.m., I saturated the as-yet dry short leachline (7.5 ft² bottom area) by filling it 6 inches above bottom with tap water. The same day, after 4 p.m., I refilled and started measuring how fast the water level was decreasing. I wanted to measure the rate of absorption when the water level was close to the bottom. At such a moment the gradient would be close to unity (the ratio of sum of surface head plus soil head, divided by soil head, would be close to unity), and the rate would be constant. The results are in Table U.2.

After these measurements, I diverted all the sewage from my septic tank to the same short leachline (with about 7.5 ft² of bottom area). It kept on receiving septic tank effluent until April 30, 1988. On this date, at noon, I cut off the flow of effluent. The level of (translucent) liquid within the short leachline went down slowly. On the following day, May 1, 1988, by 8:18 a.m., the level had gone down 1.94 inches in 18.3 hours. This is an average rate of 0.27 cm/hour or 1.6 gallons/ft²/day. I started the measurements. These are in Table U.3.

Table U.2   Drop in Water Level, Tapwater Run

Time (hr:min:sec)	Time Increment (sec)	Level (cm)	Drop (cm)	Rate (cm/hr)
4:49:22		1		
	54		1	67
4:50:16		2		
	77		1	46
4:51:33		3		
	90		1	40
4:53:03		4		
	99		1	36
4:54:42		5		
	221		2	32
4:58:21		7		
	124		1	29
5:00:25		8		
	123		1	29
5:02:28		9		
	62		0.5	29
5:03:30		9.5		

End of run after two equal measurements. Stabilized rate = 29 cm/hr. Remaining liquid level was about 4–6 cm from bottom.

Table U.3   Drop in Sewage Level, Sewage Run

Time (hr:min)	Time Increment (hr)	Level (cm)	Drop (cm)	Rate (cm/hr)
08:46		4.85		
	2.25		0.5	0.22
11:00		5.35		
	5.17		0.95	0.18
16:10		6.3		
16:11[a]	–	0.5[a]	–	–
	3.17		0.6	0.19
19:22		1.1		

End of run. Remaining liquid level was about 4–6 cm from bottom. Stabilized rate = 0.2 cm/hr.
[a]At this moment I adjusted the measurement scale so that the new level read on the scale was 0.5.

Comparing the tables' stabilized rates, one might be tempted to conclude that tap water infiltrated at an equilibrium rate 29/0.2 or almost 150 times faster. Or, that sewage infiltrated the clogging layer at a rate almost 150 times slower. This is what happened in the experiment.

Question A

Were the rates measured the actual rates of infiltration into the soil?

Answer

Yes and no. The gravel within the leachline had about 33% voids. So, the rates measured were about 3 times faster than if the gravel had been absent.

Question B

Can one conclude that (in the soil used) sewage discharged to leachlines infiltrates at a rate 1/150th as fast as pure water?

Answer

No. The experimental results cannot be extrapolated to a normal leachline with 1 to 3 feet of sewage ponded above bottom, even if the soil and temperature and microbes and sewage composition are the same. It would be like comparing apples and oranges, as explained below.

Disregarding the fact that the tap water measurements may be only one-fifth as fast as they should be because of air entrapped in the soil pores,[1] * raising the level (pressure head) of ponded water has a completely different effect from raising the level of ponded sewage. Let us disregard the fact that the wetting front is not quite the point where gradient

---

*This applies to perk test holes, too.

can be measured from (because of "capillary suction" just below the front). Let us also disregard the effect of sidewall infiltration, as it would complicate the picture even further. Let's concentrate on the bottom of the leachline trench.

Let us assume that we increase the level of ponded water from 6 to 60 cm, and that the sidewalls do not absorb any water. (This would occur in something like a 15' wide seepage bed excavated through a clay stratum until the bottom is on a loam soil stratum. If the water in the bed has been ponded for a long time, the wetting front is very deep and hence the gradient and the infiltration rate will remain almost the same, even though the ponding level increases by a factor of 10.)

Now let us vary the level of ponded sewage, after the biomat has clogged the bottom. Right below the biomat there is an unsaturated zone. For practical purposes, we can envision the wetting front as being right below the biomat and staying there even though the ponding suddenly increases from 6 to 60 cm. The biomat is like the bottom of a porous clay or fritted-glass pot. A sudden tenfold increase in the pressure head could result in an almost tenfold increase in absorption rate. Then, with time, this rate might decrease because the clogging mat enters deeper into the soil and increases the resistance (decreases gradient). The final long-term absorption rate (LTAR) might be roughly proportional to the cubic root of the ratio of ponding heights.[2] So, absorption of water and absorption of sewage are not quite comparable.

The experiment described above yielded other interesting observations. Soon after the cell dried up, I looked at the bottom of the observation pipe (a bit of geotextile with some gravel over it). Everything was covered with a film of grayish mucilage, the biomat. Two weeks later, the gravel had streaks of a white powder (possibly ascomycetes?). By March 1989 everything was as clean as the day it was installed: no traces of biomat or of white powder.

References

1. Constantz, J., et al. 1988. Air encapsulation during infiltration. *Soil Sci. Soc. Am. Proc.* 52:10–16.
2. Laak, R. 1986. *Wastewater Engineering Design for Unsewered Areas.* Technomic Publishing Co., Lancaster, PA.

PROBLEM 6. Sizing seepage pits

Question A

Can seepage pits be sized accurately on the basis of perk tests?

Answer

Usually, no.

When sizing leachlines, one can go back to much research work and even to some statistical correlation of performance versus measurement (of mpi or permeability constant). There is nothing of the kind to go back to when it comes to seepage pit sizing (unless a local contractor has kept records for 20 or more years).

This book discusses methodologies for sizing pits, and it even makes an attempt to tolerate the falling head test as currently performed in southern California. Nevertheless, I feel that it is intellectually dishonest to continue using and tolerating the current falling head methodology. The reasons are presented in the answer to question C, below.

Question B

The falling head formula indicates that the rate of water absorption in a pit or deep test hole, in total gallons per day, should be proportional to the diameter of the hole. Is this seen in the field?

Answer

Everything else being the same, double the diameter means double the absorption surface, and hence double the absorption. This should hold as long as the gradients have not decreased substantially due to water infiltration; during short-term testing, the gradients do not decrease substantially. The results of a test performed on 15'-20'-deep pits drilled with bucket augers in 1963 by Amco Engineers for Los Angeles County tracts 26263 and 27463 are consistent with the theory. (The data they reported do not prove anything because we don't know if the data were biased by the methodology and expectations of the testers. But in the absence of other information, if I had to place a bet, I would bet the ratios obtained are correct for short-term testing.) The ratios of gallons of water absorbed per day in 48"-diameter pits versus 24"-diameter pits were 1.8, 2.1, 1.6, and 2.7. Per the falling head formula, the ratio should be 2. The average and standard deviation of the reported ratios turns out to be 2.05 ±0.15.

Question C

Is there something wrong with the current falling head test for pits? If so, why is it still being used?

Answer

In the late 1970s, a committee of consulting engineers from two southern California counties analyzed the results of a series of falling head tests for sizing pits. The holes tested at the same location were about 6", 10", and 14" or more in diameter, per my rusty memory banks. The holes with diameter 10" or smaller had been drilled with screw-type flight augers; those with 14" or larger diameter were drilled with bucket augers. The measured rates of absorption per square foot of sidewall were so much higher than expected in the

14"-or-larger-diameter holes that the engineers concluded that the falling head formula Q = FD9/Lt *did not work with large-diameter holes!** So, they also concluded that the test hole diameter had to be kept below 10" and that one had to use screw augers only. It never occurred to them to look at what the screw-type augers were doing to the hole sidewalls and permeability. Imagine the sidewall as a cylinder of an adobe-like material about 1" thick with a glistening, smudged inner surface, and you'll see with your imagination what they failed to see with their eyes. Years later I pointed this out to some of them, but they replied that it was better to err on the side of safety: compacted sidewalls would yield conservative results.

Not quite. I have seen that compaction and smudging result in measurements that indicate an apparent medium or low permeability in soils that were so excessively permeable that they were unsuited for sewage disposal. (When sewage percolates down fast, it is not well filtered and it contaminates groundwater). And there are many other problems with the methodology:

- Testing screw-augered holes does not yield a rough estimate of soil permeability, but of relative soil compaction (and of the lack of concern of the tester). It is like giving an IQ test to a cadaver—we may learn something about the IQ or disposition of the tester, little about the subject of the test.
- Bucket augers, not screw augers, are the accepted type of auger per *Methods of Soil Analysis*.[1]
- About a year ago, an inexperienced county septic systems specialist forced me to use screw-augered holes. I used them, and obtained very high values. (Some Q's exceeded about 10 gallons of sewage per square foot per day. The soil might have had fractures, or portions of the com-

---

*Q is one-fifth the gallons of water absorbed per time interval t per square foot of sidewall, and is mistakenly assumed to be equal to the amount of sewage that can be applied per square foot; D is the diameter of the test hole; F is the fall in water level during the interval; and L is the average length of the water column during the interval.

pacted sidewall might have sloughed off.) The specialist suspected that there must be something wrong with the measurements because, per his calculations, a very shallow 7-foot-deep seepage pit would suffice for a 6-bedroom home. He hadn't yet learned that the experts who promulgated the falling head test had figured out a way to deal with real but absurdly high Q's, and that this way had been incorporated into southern California counties' methodology manuals. The experts had deduced that since the 1976 Uniform Plumbing Code allowed a maximum Q of 5, Q's higher than 5 must be given a value of 5. Thus, the ridiculous results of an absurd methodology were counteracted by means of a ludicrous rule!

- An accurate picture of the soil strata in which the test holes and pits are installed cannot be obtained from holes drilled by screw augers. These augers mix soil from the bottom of the hole with soil from the sides of the hole and push the mixture up and out of the hole. The soil technician, usually located at the top of the hole, has to figure out what bits of soil came from where and invent and manufacture plausible soil logs. If and when something goes wrong, and it is found that the logs aren't accurate, the consultant and boss of the technician can blame the poor technician for the screwy soil logs.

The methodology is wrong indeed! The probable reason the methodology is still being used is to be found in appendices M and O: fear of liability. Everybody uses it and has been using it for a long time, so it is legally safe to keep on doing it.

Question

Is there a better way?

Answer

If one must do falling head testing for seepage pits, most of the above-mentioned problems can be mitigated if the holes are drilled with bucket augers. The measurements obtained may be converted to "pit mpi" as shown in *Septic Systems Handbook*. Then, the same standard tables used for leachline absorption area may be used with these "pit mpi." Allow no more than 30 pit mpi. This methodology is not an accurate way to predict how much pit absorption area is needed, but it is consistent with leachline methodology and is just a bit less ridiculous than the present way of doing things.

Whatever type of auger is used to drill pit test holes, we must remember that most of the absorption might occur through a couple of barely visible fractures or root channels. So, mere test results might lead one to believe that a low-permeability soil with fractures (or root channels) is a high-permeability soil. That's why the soil profile must be accurately described (by using bucket augers) and the test results must be interpreted in light of this description. Screw augers do not result in accurate descriptions. Measuring Darcy's permeability (hydraulic conductivity) constants of some of the strata penetrated by the pit can be a very useful diagnostic tool.

It is important that jurisdictions monitor the longevity of the absorption areas installed in each locality that has distinct soil conditions. Reality in the form of records (of failure versus time) is more persuasive than theories, especially in regard to seepage pits.

Reference

1. Amoozegar, A. and A. W. Warrick. 1986. Hydraulic conductivity of saturated soils. In: A. Klute (Ed.), *Methods of Soil Analysis, Part I*, second edition, pp. 735–798. American Society of Agronomy and Soil Science Society of America, Madison, WI.

PROBLEM 7. Applicability of Darcy's formula when gradients are steep (or high)

In some engineering and hydrogeology books, one finds that k (Darcy's permeability constant or hydraulic conductivity) appears in formulas in which the hydraulic gradient is expressed as the differential of the variable in the ordinate (dh), over the differential of the variable in the abscissa (dx). The derivation of the classical well formula used to predict Q (water flow into wells) assumes such a dh/dx gradient. So does this book (see Chapter 13, Section 13.1.2, formula 14).

Questions

What's wrong with such formulations? Is there something wrong with the Darcy formula Q/S = k dh/dx? Why do some professionals believe that Darcy's formula doesn't work well when the gradient exceeds an angle of 45°?

Answers

In analytic geometry or calculus, the ratio, "differential of the ordinate variable (dy or dh) over the differential of the abscissa variable (dx)," corresponds to the tangent at a point in the curve of y (or h) versus x. The value of the tangent varies from zero (0°, horizontal) to infinity (90° to the horizontal, vertical angle).

Per Darcy's formula, $Q/S = k \sin\theta$

The sine of an angle varies from zero (at 0°) to 1 (at 90°). The constant k is measured when the flow is vertical (90°); then the sine is equal to 1.

Now, most natural bodies of water move under a gradient or slope of just a few degrees and the sine and the tangent are numerically quite similar, so usually it doesn't matter too much whether the Darcy formula used is the true one (the

Table U.4   Angles, Slopes, and Values of Tangents and Sines[a]

Angle (degrees)	Slope	Tangent	Sine
1	0.017	0.017	0.017
5	0.087	0.087	0.087
10	0.17	0.17	0.17
20	0.36	0.30	0.34
30	0.57	0.57	0.50
45	1.00	1.00	0.71
60	1.73	1.73	0.87
89	57.0	57.0	0.99
90	infinity	infinity	1.00

[a](Note that the values given for the slope, when multiplied by 100, give the more familiar values of "percent slope.")

one with the sine), or the one with the differentials; but at or beyond the 45° angle (100% slope) the values of the sine and the tangent diverge progressively and significantly. (This is the main point I had in mind when I wrote question No. 5 in Appendix L.) See Table U.4.

PROBLEM 8. Filter to control phosphate and nitrate

An interesting way to reduce the discharge of phosphate and nitrate has been developed in Canada. Septic tank effluent is discharged over a large sand filter. Within the sand filter, there is a horizontal layer of sand mixed with red mud (a by-product of bauxite manufacture that contains oxides of iron, aluminum, and calcium) to trap phosphate and also another layer of sand mixed with a zeolite (clinoptilolite) to trap ammonium. Eight inches of a layer made with 96% sand plus 4% red mud captures phosphate, reduces it from about 14 mg/L to about 5 mg/L, and allegedly lasts 50 years under a load of 1 gallon/ft^2/day. The sand with clinoptilolite above this layer reduces the total nitrogen concentration from about 36 mg/L to 7 mg/L.

Clinoptilolite selectively captures ammonium (and potassium) ions because these fit almost exactly within its crystal

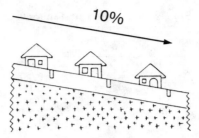

Figure U.1 Cross section of upslope development.

lattice holes (negatively charged hexagonal holes). About 3.3 g of ammonium ion can be captured by 100 g of clinoptilolite.

Question

If a home will discharge 14 kg N per year, what weight of clinoptilolite would be required to trap this yearly discharge? Assume 100% efficiency.

Answer

Per the periodic table of the elements, 14 kg N are equivalent to 18 kg NH_4^+. If 3.3 g NH_4^+ require 100 g of clinoptilolite, then 18 kg will require 540 kg of clinoptilolite per year.

PROBLEM 9. Transmission of percolates

Visualize a subdivision on a 10% slope, as shown in Figure U.1. The dimension of the subdivision along the slope is 500 feet, west to east; the dimension perpendicular to the slope, north to south, is 1225 feet. The soil is dry and 25 feet thick. Below it, the bedrock is impermeable. The k of the soil is 2.6 gallons/ft²/day, or, what's the same, 0.35 feet/day. (2.6 gallons/ft²/day divided by 7.5 gallons/ft³ gives 0.35 feet/day). If the subdivision is developed as planned, each day 18,000 gallons of sewage will be discharged through leachlines.

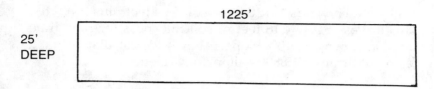

$$1225 \times 25 = 30{,}625 \ FT^2$$

Figure U.2 Cross section of soil stratum.

Assume there is no rainfall and no irrigation.

Question A

How many gallons/day can the site transmit down the 10% slope before sewage surfaces on top of the ground?

Answer

The cross section perpendicular to the direction of flow is $25 \times 1225 = 30{,}625 \ ft^2$, as represented in Figure U.2. The sine of a 10% slope is 0.1. Hence, $Q = 30{,}625 \times 2.6 \times 0.1 = 8800$ gallons/day can be transmitted.

Question B

How many gallons of sewage could the site absorb and transmit if leachline bottoms are at a vertical depth of 4 feet and one needs 5 feet of unsaturated soil below the leachlines?

Answer

The projection of the 4 feet of vertical depth onto the cross section perpendicular to flow is 4 feet times the cosine of the angle, which yields 4' × 0.99 = still 4'. 4 feet plus 5 = 9 feet. This leaves 25−9 = 16 feet of soil thickness for flow transmission down the 10% slope. Above, we calculated that 25 feet could transmit 8800 gallons/day. Hence

$$25/8800 = 16/X$$

and solving, X = 8880 × 16/25 = 5600 gallons can be safely absorbed and transmitted.

PROBLEM 10. Transmission of percolates in dipping strata

The site of a proposed subdivision is fairly horizontal and flat, and measures 500 by 1225 feet. The soil strata dip at a 40° angle. Figure U.3 shows a vertical cross section along (and parallel to) the 500-foot dimension; note the angle of dip of the strata and how the seepage pit perforates the strata. The strata are sandstone interbedded with impermeable shale. The shale occupies one-third of the profile; sandstone occupies two-thirds. The k of the sandstone is 2.6 gallons/day.

The proposed discharge of sewage will be 18,000 gallons/day. Assume that the top of an aquifer below is so deep that there will be no interference. Also assume that there is no rainfall and no irrigation.

Question A

If enough pits are installed throughout the site so that the site can absorb the proposed discharge, will all of the absorbed sewage flow (laterally) down? The pits are 25 feet deep below the inlet.

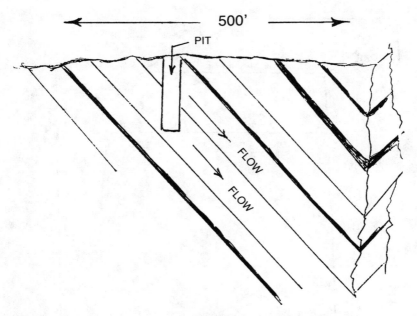

Figure U.3   Flow from a seepage pit into dipping strata.

Answer

Refer to Figure U.4. Below the bottom of the pits, the thickness of the cross-sectional area in which flow occurs (this cross section is perpendicular to flow) is 500 feet times sine of 400, or 320 feet; the length of this area is 1225 feet. The area perpendicular to flow below the bottom of the proposed pits is then $320 \times 1225 = 392,000$ ft². If the pits are 25 feet deep below the inlet, then the projection of 25 feet onto the perpendicular to flow is 25 cos 400 = 20 feet. Then $320 + 20 = 340$ feet, and the total area perpendicular to flow is $340 \times 1225 = 416,500$ ft².

Then, $Q = S \, k \, \sin\theta = 416,500 \times 2.6 \times 0.64 = 693,000$ gallons/day.

But sandstone occupies only 2/3 of the cross section per-

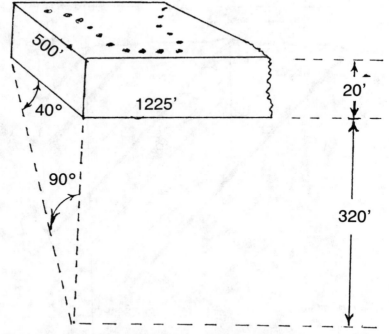

Figure U.4   Location of seepage pits (dots). *Note:* Graph is not to scale.

pendicular to flow. Hence, 2/3 of the above gallonage is 462,000 gallons/day.

Therefore, the strata can transmit much more than 18,000 gallons.

Question B

Refer to the situation above, but imagine that the sewage discharged is going to be 400,000 gallons per day. Imagine that the top of the aquifer is 20 feet below the depth of the pits. Would you be concerned?

Answer

I would. When the discharge is close to the maximum saturated flow that the strata can transmit, much of the flow will be saturated (just above each impermeable stratum), and pathogens may not be removed efficiently before the flow reaches the aquifer. Furthermore, mounding might elevate the top of the aquifer all the way up to the seepage pits.

One might also wonder how much aerobic decomposition of the organic sewage constituents would occur deep in a stratified soil with very poor permeability to air.

PROBLEM 11. Nitrate impact and well depth

A long, narrow valley has an aquifer moving throughout its length, from east to west. The width of the valley and of the aquifer is 3000 feet (north to south), and the wetted depth of the aquifer is 200 feet. A proposed subdivision tract will spread through the northerly one-third of the 3000-foot width and will discharge 690 kg N in the form of nitrates through its septic systems. The soil is so sandy that no significant denitrification will take place. The water in the aquifer contains almost zero ppm (mg/L) of N and can accept 10 ppm.

The minimum flow through a vertical north-south cross section of the aquifer is 950 acre-feet per year, moving east to west. Assume that all rainfall is lost to evapotranspiration.

Question A

If each lot owner is going to have his or her own well, how deep should the well pumps go below the top of the aquifer in order not to pump water with more than 10 ppm N? Assume that nitrogen-containing plumes are diluted uniformly in the horizontal plane, north to south.

Answer

The wetted cross section perpendicular to the aquifer flow is $200 \times 3000 = 600{,}000$ ft^2. 950 acre-feet contain $950 \times 43{,}500 = 41{,}000{,}000$ ft^3. Therefore, the Darcy velocity of flow is $41{,}000{,}000/600{,}000 = 69$ feet per year, east to west.

Now the problem is reduced to finding the depth of a parallelepiped of water (measuring 1000 feet north to south and 69 feet east to west) sufficient to dilute the N to 10 ppm. Visualize this parallelepiped moving below the tract, entering the tract on the tract's east side (with a concentration of 0 ppm N) and coming out of the tract's west side (with an average concentration of 10 ppm N) every year.

Each foot of depth of the parallelepiped has 69,000 ft^3 and weighs 69,000 ft^3 $\times$ 28 kg/ft^3 = 1.9 million kg.

To dilute the N we need

$$10 \ (\text{parts})/1{,}000{,}000 \ (\text{parts}) = 690 \ (\text{kg N})/X \ (\text{kg N})$$

Solving for X, X = 69 million kg water.

Dividing 69 million by 1.9 million per foot of depth we get 36 feet of depth.

(I'd start the perforated casing at 20 or 30 feet depth and then I'd add the minimum of 36 feet. The additional 20 or 30 feet of depth would help to minimize the danger of pumping too close to a poorly diluted nitrate plume, near the top of the aquifer.) Note that the 36 feet of depth refers to the west side of the tract. The minimum depth of wells at the east side is zero. Elsewhere, depth can be obtained by simple interpolation.

Question B

If the background N content of the aquifer had been 1 ppm, the minimum depth of the pump would change to Y feet. What's the value of Y?

Answer

Multiply 36 times the ratio $10/(10-X)$, where X is the initial N concentration in ppm. If $X = 1$, Y is 40 feet. *One part per million (ppm) is virtually the same as one milligram per liter (mg/L).*

PROBLEM 12. Nitrate impact per the "cube" or parallelepiped method

Assume that (1) all of the nitrate generated in a proposed subdivision reaches the aquifer below it; (2) aquifer water is extracted with a hypothetical well that reaches down to the bottom of the aquifer; and (3) the nitrate is mixed and diluted by the water pumped all the way down to the depth of the well. If after this mixing the concentration increase is 1 ppm N or lower, the jurisdiction will approve the subdivision.

See if the following tract can be approved. Use the given data and assumptions.

Tract XXXXX measures 1320 feet east to west and 660 feet north to south. It has 70 single-home lots. The soil is very sandy. Data provided by reputable hydrogeologists give basic information: Darcy velocity = 0.24 feet/day moving in a 239° southwest direction; aquifer saturated thickness = 938 feet. The number of residents is 269. Yards (lawns) will occupy 567,000 ft², or 5.25 ha.

Conservatively, assume that all of the 6 kg of nitrogen excreted per person per year reach and mix with the portion of the aquifer penetrated by the wells. (Actually, in this case 10–20% may be lost via denitrification and septage pumpings, and most of the N will remain in upper aquifer strata for many miles of travel.) The dilution provided by the about 3 to 4 feet/year of lawn irrigation percolate plus septage that percolates down to the aquifer is negligible, since the aquifer wetted depth exceeds 900 feet. (Note that since the following calculations use the Darcy k, the 900 feet of depth is treated as that of a "lake" of pure water: for all practical purposes, the soil that holds the water has zero volume or depth. The rough estimate of 3–4 feet per year of irrigation percolate is based on pan evaporation of 5–6 feet per year times 0.80 equals evapotranspiration, and 50% of the irrigation is leached down as percolate. For more precision, one should consult the local Agricultural Extension Service. The feet of sewage percolate are calculated by converting the expected discharge of sewage to acre-feet of sewage and dividing by the acreage of the tract.) Assume that the best approximation available for the amount of N in lawn leachate derives from the data in T. Morton et al. ("Influence of overwatering and fertilization on N losses from home lawns," *J. Environ. Qual.* 17:124–130, 1988). (A permeable, overwatered and over-fertilized Long Island soil leached up to 32 kg N per ha per year.)

CALCULATIONS for Tract XXXXX

1. Volume of dilution
   a. Wetted depth of aquifer is 938′.
   b. Since the flow is in a direction 239° toward the southwest, the width of the tract projected onto the perpendicular to the direction of flow is about 1320′ cos 59° = 680′. (Note that 239° − 180° = 59°.) See Figures U.5(a) and U.5(b).
   c. Darcy velocity is 0.24 feet/day.
   d. 1 m³ equals 33 ft³; 1 year equals 365 days.

Hence, 938 × 680 × 0.24 × 365/33 = 1.69 million m³/yr
$$= 1690 \text{ million kg/yr}$$

2. Amount of N
   a. At 6 kg N/person/year, 269 people discharge 1614 kg N/year. Q
   b. 5.25 ha of lawns leach down a maximum of 168 kg N/year. Q
   c. Total maximum N = 1782 kg N/year (8000 kg of nitrate/year). Q

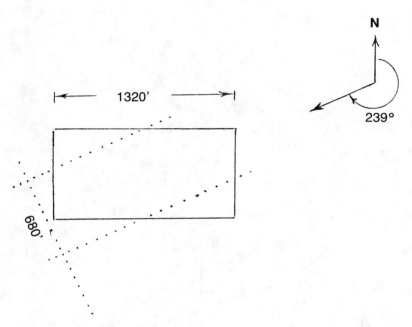

Figure U.5(a) Bird's-eye view of the nitrate plume moving in the direction of flow. Flow is 239° southwest.

3. Increase in concentration
   1782 kg N/1690 million kg water = 1.0 ppm N (4.74 ppm nitrate). (One part per million, ppm, is virtually the same as one milligram per liter, mg/L)

PROBLEM 13. Lot size and cost effectiveness of pollution control strategies

There are many strategies to reduce septic systems' nitrate impact on groundwater. One strategy is to control lot size. This strategy is very involved and is still not well developed. Another strategy is to decrease the nitrate output of septic systems with devices such as recirculating sand filters, batch nitrification/denitrification, dosing regimes, etc.

0.24 ft/day = 87 ft/yr

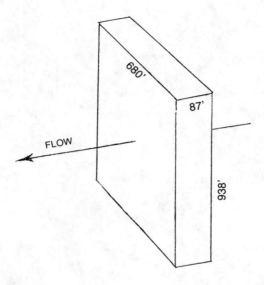

Figure U.5(b) Dimensions of the water parallelepiped.

To my knowledge, where individual septic systems are used, the most simple and elegant device requiring minimum maintenance, is the experimental RUCK system.[1] It is very expensive, but it might remove up to 90% of the nitrogen if all conditions are optimal (which is not always the case). A 90% removal would come close to reducing the nitrate in the percolate to about the concentration in drinking water standards, 45 mg/L of nitrate. In an independent test by Lamb et al.[2] a RUCK system removed only about 50% of the nitrogen. However, efficiency is not the only criterion to be considered. Benefit/risk and benefit/cost analyses are the important criteria.

If a local jurisdiction determines that the benefit of allowing an area to develop (on septic systems with innovative nitrogen reduction devices) exceeds the risks and costs involved, the developer and his consultant may use simple benefit/cost rules to determine if the project is feasible. For instance, assuming that a given innovative system will remove not less than 40% of the nitrogen, would it be more cost effective to install that system at $10,000-$15,000 per house or to sewer the development? Would the development still leave the expected margin of profit to make it worthwhile?

Question

Determine how many innovative systems have to be installed in a 40-lot tract with 15,000-ft^2 lots. The local jurisdiction requires 20,000 ft^2 per lot. The jurisdiction asserts that 20,000-ft^2 lots do not result in groundwater degradation by nitrates. So, calculate how many innovative systems would be required to reduce the impact of the tract to the level it would have had if its lots were 20,000 ft^2; assume the innovative system is capable of removing 40% of the nitrogen (in the septic effluent). Since the soils are sandy, assume 0 to 10% nitrogen removal in the soil, as reported elsewhere.[3]

Answer

Let "A" be the amount of nitrate produced by 40 dwelling units, each in a 20,000-ft^2 lot. Let the average amount of denitrification be $(0 + 10\%)/2 = 5\%$. And $1 - 0.05 = 0.95$. Then, the relative amount of nitrate produced on a per-unit-area basis in a normal tract with 20,000-ft^2 lots is

$$(40 \times 0.95 \text{ A})/20,000 = 0.0019A$$

Similarly, the relative amount of nitrate generated by 40 15,000-ft^2 lots consisting of X lots with innovative system (and 40% N loss) plus Y lots with no system (and 5% N loss) is

$$(X \times 0.6 \text{ A})/15,000 + (Y \times 0.95 \text{ A})/15,000$$

Equating the two previous relationships and simplifying, we obtain Equation 1:

$$0.6X + 0.95Y = 15,000 \times 0.0019 = 28.5 \qquad (1)$$

and also we have that

$$X + Y = 40 \text{ (lots)} \qquad (2)$$

Solving for X and Y in the two linear equations above, we obtain

$$Y = 13 \text{ lots with no system}$$

$$X = 27 \text{ lots with system}$$

Thus, there should be no significant difference between the extent of nitrate generated by 40 dwelling units in 20,000-ft^2 lots versus 40 dwelling units in 15,000-ft^2 lots, provided that 27 out of the latter 40 dwelling units are served by the innovative system.

References

1. Laak, R. 1987. RUCK systems. Civil Engineering Department, University of Connecticut, Storrs, CT.
2. Lamb, B., et al. 1987. Evaluation of nitrogen removal systems for on-site sewage disposal. In Proceedings of the Fifth National Symposium on Individual and Small Community Systems. ASAE Pub. 10–87, pp. 151–160. American Society of Agricultural Engineers, St. Joseph, MI.

3.   Broadbent, F. E., and H. Reisenauer. 1984. Fate of waste-
     water constituents in soil and groundwater. In "Irriga-
     tion with Reclaimed Municipal Wastewater." Report No.
     84–1wr. California Water Resources Control Board.

PROBLEM 14. More complex calculations about nitrate pollution

Question

Show that the nitrogen concentration in the upper 100 feet
of an aquifer below a subdivision tract will not exceed 1 ppm
N averaged over the top 100 feet of the aquifer. The data are
as follows:

- The average 3-bedroom home has 3 occupants and generates 300 gallons of sewage per day; in this tract it will
  be assumed to generate $300 \times (3.3/3) = 330$ gallons per
  day.
- The tract has 60 lots in 13 acres.
- The N generated by 60 homes with 3.3 people each at 6
  kg N/capita/year is 1190 kg.

The amount of sewage generated by these homes at 100
gallons per capita per day is $60 \times 3.3 \times 100 \times 365 =$
7,227,000 gallons per year, or 970,000 ft^3 per year, or 1.7 feet
per year over the 13 acres. With at least 1 foot of lawn irrigation water leachate, the total percolation flow adds up to
about 3 feet per year, or a bit more. Assume that the irrigation leachate contains a negligible concentration of N.

A reputable hydrogeologist has provided the following: a)
data to the effect that the aquifer moves at a velocity of 0.38
feet/day and b) a plate of mounding produced by 3 feet of
percolate per year; as seen in Figure U.6, the percolate
mounds and spreads out over the aquifer with a radius
exceeding 6600 feet. (This is per a computer program based
on Hantush's formulas published in 1967 in *Water Resources*

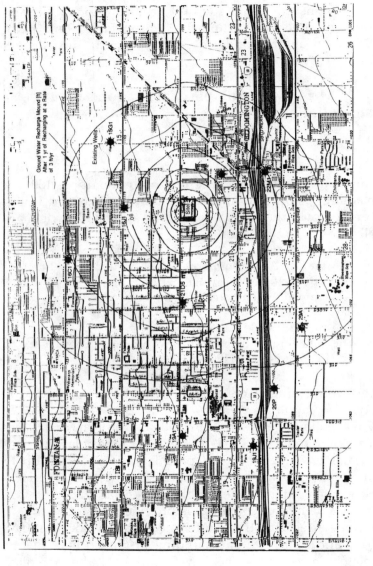

Figure U.6   Groundwater recharge mound after one year of recharging at a rate of 3 ft/yr. Groundwater depth is in excess of 300 ft.

Research 3:227–234.) The aquifer has a thin "veneer" of high-N water on top.

Assume that the distribution of N over the surface of the aquifer is proportional to the height of the percolate mound over it. If the calculations were to average all the N input over the whole spread of the mound, this would result in two problems:

1. The "tails" at the tips of the mound vertical cross section are almost asymptotic. A very thin veneer extends over a very large area. The average N content would appear to be very low once it is averaged over the large area.
2. At the same time, in the aquifer closer to the center of the mound, the concentration of N would be much higher than the gross average.

Procedure and answer

Calculate the radius at which the volume of the mound is 50% of the total volume. This volume contains 50% of the N. If this higher-than-average N concentration (which is mixed or diluted over a small aquifer cross section) meets the 1 ppm nitrogen criterion set by the jurisdiction, the average (after dilution over a large aquifer cross section exceeding 6600 feet in radius) must meet it too, since its concentration is bound to be lower.

First plot the heights of the mound and their locations; these can be obtained directly from the "circles" in Figure U.7. In this plot construct graphically a cone (dashed lines) with height = 0.18 = 0.2 feet and radius = 6600 feet. (See Figure U.8.) Calculate the relative volume of this cone, and then calculate the radius of the inner part of the cone, which yields half the full volume of the cone. This turns out to be about 3350 feet from the center.

The calculations for the volume V of the cone in Figure U.8. are below.

Let X = radius, and Y = height of the cone

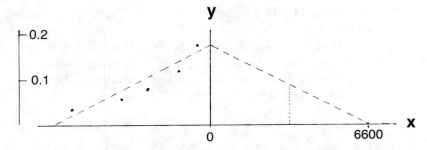

Figure U.7 Mound dimensions in feet. *Note:* Graph is not to scale.

$$V = 2\pi \int_{x=0}^{6000} XYdX$$

and

$$Y = 0.2 - 0.2X/6600$$

Substituting the value of Y, we have that

$$V = 2\pi \int_{x=0}^{6600} X(0.2 - 0.2X/6600)dX = 2\pi 1.5 \times 10^6$$

Calculations for the value of X, which encompasses 50% of the volume of the cone:

$$0.5(2\pi 1.5 \times 10^6) = 2\pi \int_{x=0}^{x} X(0.2 - 0.2x/6600)dX$$

$$1.5 \times 10^6 = 2[(0.2X^2/2) - (0.2X^3/19,800)]$$

Simplifying both sides,

$$14,850 \times 10^6 = X^2(1980 - 0.2X)$$

Solving for X, X turns out to be about 3350 feet.
Now, the more traditional methodology. We have a paral-

lelepiped 6700 feet wide (3350' × 2), 100 feet deep (per county requirement), moving at 0.38'/day or 137'/year. Over this body of water is half of the N input. This body of water contains 6700 × 100×137 = 92,000,000 ft³. (A cubic foot of water weighs 28.3 kg.) Hence, this body of water weighs 2600 million kg.

Half of the N discharged in the tract weighs 1200/2 = 600 kg. Therefore, once diluted, the maximum average increase in concentration near the center of the mound is

$$600/2600 = 0.2 \text{ ppm N } (0.9 \text{ ppm nitrate})$$

(By looking at Figure U.8, we can estimate that the average height of the cone above the Y, which corresponds to X = 3350, is between roughly 0.08 and 0.2, say 0.14 ±50%. Hence, at the very center of the cone, the maximum increase in nitrogen concentration should be 0.2 ppm N plus 50% or 0.3 ppm, which is substantially less than the 1 ppm criterion.)

PROBLEM 15. Inadequacy of Uniform Plumbing Code sewage disposal rates

The requirements of the Uniform Plumbing Code (1976–1988 editions) for leaching areas are confusing, contradictory, and designed to ensure short longevity.

For instance, in the case of fine sand, the UPC Table I-4 requires 25 ft² of leaching area "per 100 gallons" of who knows what. Since the same table assigns fine sand a maximum absorption capacity of 4 gallons/ft², and 100/25 = 4, one may conclude that "per 100 gallons" refers to actual flow of sewage effluent. The same conclusion can be arrived at by comparing the respective figures for other soil textures in UPC Table I-4. But in UPC Table I-5, one finds that the requirement is defined as "per 100 gallons of Septic Tank Capacity." A former IAPMO Board president assured me that this is the case for Table I-4 as well; and this same interpretation is given by most if not all jurisdictions!

Assume the following situation: three single homes will be built. Their leachlines will be installed in fine sand. Home A will have two residents and two bedrooms. Home B will have 4 residents and 4 bedrooms. And home C will have 6 residents and 6 bedrooms.

Question A

Calculate the amount of absorption area per resident according to UPC requirements. (Note that residents do generate sewage. Homes, bedrooms, or septic tanks of given sizes do not generate sewage.)

Answer

Given the number of bedrooms, the UPC requires the following septic tank sizes: home A, 750 gallons; home B, 1200 gallons; and home C, 1500 gallons.

Hence, at 25 ft² per 100 gallons of tank capacity, homes A, B, and C will have, respectively, the following square feet of leachline absorption area:

$$25 \times 7.5 = 188$$

$$25 \times 12 = 300$$

$$25 \times 15 = 375$$

Therefore, the per capita square feet of absorption areas are: home A, 94; home B, 75; and home C, 62.

From the above, it is evident which septic system will be more prone to septic system failure.

Question B

The UPC assumes that a bedroom generates 150 gallons of sewage per day. If this figure is correct, how many gallons per day per square foot of absorption area are discharged in the leachlines serving homes A, B, and C?

Answer

$$A: \quad 2 \times 150/188 = 1.6$$
$$B: \quad 4 \times 150/300 = 2$$
$$C: \quad 6 \times 150/375 = 2.4$$

Question C

Winneberger, Laak, Machmeyer, Bouma, the EPA . . . just about everybody but the UPC recommends a maximum sewage disposal rate of about 0.8–1.2 gallons/ft²/day for sand. The 0.8-gallon figure is more correct because the 1.2-gallon figure refers to sewage applied in about 4 doses per day, and this usually does not occur in most installations. Compare the 0.8-gallon figure to the 1.6 to 2.4 gallons/ft²/day derived in Question B. Would you expect the septic systems in houses A, B, and C to suffer premature failure?

Answer

I certainly would!

Question D

Would the answer to Question C be different if the soil were not sand?

Answer

No. The worst type of soil, "clay with small amount of gravel," is assigned a disposal rate of 0.83 gallons/ft²/day per the 1988 edition of the UPC. But, per the authorities mentioned in Question C, this type of soil should receive roughly about 0.4 to 0.2 gallons/ft^2/day or less. The UPC allows excessive loading in all types of soil.

Question E

If one repeats the calculations for questions A, B, and C above, and this time one uses UPC rates "per 100 gallons of actual flow" instead of "per 100 gallons of septic tank capacity," would any of the answers change?

Answer

Yes. The UPC requirements would look even worse, because the UPC rates would allow even more gallonage to be discharged per square foot of absorption area.

Question F

If you were a county septic systems specialist, would you allow designs based on the UPC sewage application rates?

Answer

If I had the power to dictate application rates, I'd answer with an emphatic "No." Please note that the UPC gives minimum requirements, and that it allows you to impose more stringent requirements.

Question G

If you are a septic systems consultant, would you be liable for a premature septic system failure if your design application rate was based on the UPC?

Answer

The courts will tell. Legally, overall, the UPC is a good "security blanket." But by now it is public knowledge that the UPC rates are very inadequate, and a court might find that you were negligent if you used the UPC application rates.

PROBLEM 16. Purely legal stuff

Within your county, there is a community where each home has its own well and septic system. One of the members of this community is being taken to court because he has been using a septic system with leachlines that violate the minimum county requirements regarding the separation between the bottom of the leachlines and groundwater, and he has refused to make corrective changes. You are to be the expert witness for the plaintiff (for your county, or for the defendant's neighbor).

A crucial point in law is that the plaintiff has to show that the defendant's actions or inactions are causing harm. You have no data regarding excess morbidity or mortality near the defendant's property. You don't even have data showing amounts of hazardous microbes in wells near the defendant's septic system. The neighboring wells are at least 1000 feet away from the problematic septic system, but the soils are so coarse that there is a possibility of microbial contamination.

Question

During your deposition, the defendant's attorney asks you: "Can you prove that my client has caused any harm? Do you have data on how many people got sick after drinking well water? Did you have the water analyzed?"

What would your answer be? Give it some thought.

Hint: Put the ball in your court.

Suggested Answer

The following answer has made plaintiff's attorneys very happy:

> Your questions are irrelevant. One can drive fast across a street intersection while the red light is on, and not kill anyone, but that's not the point. The point is that eventually someone might get killed, and that's why traffic laws must be obeyed. Your client's septic system might or might not harm others, but it violates the minimum code requirements. If everyone would be allowed to violate these requirements, some septic systems would cause problems, some would not. The only way to ensure safety for all is to ensure that everyone obeys the law or the code.

Glossary

aerobic medium with oxygen readily available for microbial metabolism

alternative system any system of sewage disposal other than conventional sewers

anaerobic medium having little or no oxygen available for microbial metabolism

ASTM American Society for Testing Materials

bedrock the rock underlying soils

BOD Biochemical Oxygen Demand; the amount of oxygen consumed by sewage microbes under standard test conditions. BOD_5 means BOD measured after 5–day incubation period

caliche a layer of soil cemented by precipitated calcium and/or magnesium carbonate, either in nodules or massive; clay and precipitated gypsum are also called caliche, though not accurately

clay (mineral) a natural soil crystalline inorganic mineral formed by decomposition of or synthesis from other minerals, and less than 0.002 mm in diameter

clay (particle) any inorganic soil particle less than 0.002 mm in diameter

clay (soil) soil with $>$ 40% clay, and $<$ 45% sand and $<$ 40% silt, on a weight basis

COD Chemical Oxygen Demand; amount of oxygen consumed when sewage is oxidized by chemical means

denitrification biochemical reduction of nitrate or nitrite to gaseous nitrogen or nitrogen oxides

distribution box   a box that receives sewage and allows it to escape through level outlets of the same size so that the sewage flow is split into equal flows

drop box   a modified distribution box; it discharges effluent only when the leachfield that discharges into it is full

exchange capacity   the total ionic charge of the soils that is active in the adsorption of cations or anions

EIS   Environmental Impact Statement; a comprehensive study and report that must be prepared under the U.S. Environmental Quality Act for projects that might have a significant impact on the environment

igneous rock   rock formed by the cooling of molten magma, rock which has not changed much since it formed

incidence   the number of occurrences in a given population during a given time period

infiltration   movement of liquid through a surface

invert   a pipe like an upside-down J that directs overflow from a full leachfield to a lower-lying leachfield

isotropic   having the same properties in all directions. As used in this book, "isotropic" should be understood to mean "isotropic and homogeneous"

loam   soil containing 7% to 27% clay, 28% to 50% silt, and < 52% sand

mound   a protuberance of water over a layer of water, or over a low-permeability stratum, or a protuberance of imported soil materials over soil (soil mound)

morbidity   incidence of illness

mortality   prevalence (percent) of deaths in a population

pathogen   any microorganism that can cause disease

percolation   movement of liquid through a porous body in response to gravity

perk rate   rate of water level fall in a test hole under standard conditions

perk time   the time it takes for the water level in a test hole to drop one inch

perking   word that substitutes for the misnomer "percolation testing"

permeability   the relative ease with which gases or liquids penetrate through the soil

pore space   the volume not occupied by soil particles in a bulk volume of dry soil (also, "voids")

prevalence   the percent of a population that is affected at a given time period

replacement area   see *reserve area*

reserve area   area in a lot that is reserved for future leach-fields in case the original ones fail

sand (particle)   a soil particle between 0.05 mm and 2 mm in diameter

sand (soil)   soil with more than 85% sand, and in which percentage of silt plus 1.5 times percentage of clay does not exceed 15

sesquioxides   oxides with formula M_2O_3, where M is a metal, commonly iron or aluminum

silt (particle)   a soil particle between 0.002 mm and 0.05 mm in diameter

silt (soil)   soil with 80% or more of silt and < 12% clay

step-down   a vertical barrier or "plug" at the end of a leach-line, with a hole near the top, that allows overflow to fall to a lower leachline

structure   the type of arrangement of soil particles

TDS   Total Dissolved Solids, or "salts"

texture   the relative proportion of sand, silt, and clay in a soil

U trap   a U-shaped pipe below any sink or drain that holds water and forms a seal so that gases from sewers or septic tanks cannot escape up the sink or drain (the correct trade name is "P trap")

vector   in biology, a living organism or inanimate object capable of transmitting pathogens

Index